D1165760

Between the Absolute
and the Arbitrary

Between
THE
Absolute
AND THE
Arbitrary

C ATHERINE Z. E LGIN

Cornell University Press

ITHACA AND LONDON

First published 1997 by Cornell University Press
First printing, Cornell Paperbacks, 1997

Printed in the United States of America

This book is printed on Lyons Falls Turin Book, a paper that is totally chlorine-free and acid-free.

Library of Congress Cataloging-in-Publication Data

Elgin, Catherine Z., b. 1948
  Between the absolute and the arbitrary / Catherine Z. Elgin.
     p.   cm.
  Includes bibliographical references and index.
  ISBN 0-8014-3359-2 (cloth : alk. paper).—ISBN 0-8014-8399-9
(pbk. : alk. paper)
  1. Philosophy.   2. Knowledge, Theory of.   I. Title.
B945.E413.B48   1997
121—dc21   96-444469

Cloth printing   10   9   8   7   6   5   4   3   2   1
Paperback printing   10   9   8   7   6   5   4   3   2   1

For Nelson Goodman

Contents

Acknowledgments

The papers that make up this volume were written over many years. In the writing I have incurred numerous debts. The critical acuity of Jonathan Adler, Nelson Goodman, and Israel Scheffler should be evident throughout. Mary Kate McGowan, Ruth Anna Putnam, and Jamie Tappenden gave me excellent advice about Chapter 1. Jill Sigman's comments on Chapter 5 were invaluable. W. V. Quine showed me how to avoid embarrassing errors in Chapter 6. Bill Lycan made useful comments on Chapter 11. Carmo d'Orey's insights into Goodman's philosophy of art enriched my understanding of both Goodman and art. Chapter 3 reflects some of what I learned from her. Allison Crapo reviewed the entire manuscript and provided much needed editorial assistance. My husband, Jim, and my son, Sam, encouraged me in, and distracted me from, my work. I am grateful to you all, and to the others whose contributions I have, no doubt, forgotten to record.

I am also grateful for the financial support I received from a variety of institutions: The National Endowment for the Humanities, The Mary Ingraham Bunting Institute, The Radcliffe Research Partners Program, The American Council of Learned Societies, The John Dewey Foundation, and The University of North Carolina Research Council.

Most of the essays in this volume have been previously published. Their original sources are as follows:

"Unnatural Science," *Journal of Philosophy* 92 (1995): 289–302.

"Theory Reduction: A Question of Fact or a Question of Value?" in *Physical Sciences and the History of Physics, Boston Studies in the Philosophy of Science* 87, ed. Robert Cohen and Marx Wartofsky (Dordrecht: Reidel, 1983), 75–92. © 1984 by D. Reidel Publishing Company. Reprinted by permission of Kluwer Academic Publishers.

"Relocating Aesthetics," *Revue Internationale de Philosophie* 46 (1993): 171–186.

"Facts That Don't Matter," in *Meaning and Method: Essays in Honor of Hilary Putnam*, ed. George Boolos (Cambridge: Cambridge University Press, 1990), 17–30. Reprinted with the permission of Cambridge University Press.

"Restoration and Work Identity," in *Representation: Relationship between Language and Image*, ed. S. Levialdi and C. Bernardelli (Singapore: World Scientific, 1994), 101–107. Reprinted with the permission of World Scientific Publishing Co.

"Translucent Belief," *Journal of Philosophy* 82 (1985): 74–91.

"Mainsprings of Metaphor" (with Israel Scheffler), *Journal of Philosophy* 84 (1987): 331–335. Reprinted with permission of the coauthor and the publisher.

"Index and Icon Revisited," in *Peirce's Doctrine of Signs: Theory, Applications, and Connections*, ed. Vincent Colepietro and Thomas Olshewsky (The Hague: Mouton, 1995), 183–192. Reprinted with permission of Mouton de Gruyter.

"Sign, Symbol, and System," *Journal of Aesthetic Education* 25 (1991): 11–21.

"The Relativity of Fact and the Objectivity of Value," in *Relativism: Interpretation and Confrontation*, ed. Michael Krausz (Notre Dame: University of Notre Dame Press, 1989), 86–98.

C. Z. E.

*Between the Absolute
and the Arbitrary*

Disturbances in the Field

The alternatives are stark. Unless answers to philosophical questions are absolute, they are arbitrary. Unless a position is grounded in agent-neutral, determinate facts, it is right only relative to a perspective that cannot in the end be justified. Philosophers differ over where the line is to be drawn. Bernard Williams takes it to divide theoretical from practical reason. Ian Hacking draws it between the natural and the social sciences. Richard Rorty believes that every branch of inquiry falls on the arbitrary side of the divide. Tom Nagel's *View from Nowhere* plots a more circuitous route. To be sure, not everyone accepts such a picture. But even those who reject the dichotomy often frame their positions in terms of it, factoring complex theoretical constructions into absolute and arbitrary components. This bipolar disorder incapacitates philosophy, preventing it from seeing how fact and value intertwine, where art and science intersect, how human agents contrive categories, set standards, define goals, and thereby fix the frameworks within which objective judgments can be made.

We may understand our predicament better if we recall the way it arose. The logical positivists undertook the admirable task of eliminating nonsense from the cognitive realm.[1] Within the realm of the cognitive, they believed, matters could be settled absolutely, once and for all. Outside, all was arbitrary. Convinced that nonsense is a linguistic phenomenon, they attempted to contrive a criterion that would distinguish between significant and nonsensical locutions. Although they refined the criterion over the years, its central features were retained.[2] Only two sorts of claims were considered cognitively significant: those whose truth values are entirely determined by their logical forms and those whose truth values can be established by empirical evidence. The former belong to logic or mathematics, the latter to empirical science.

The scope of empirical science, as the positivists used the term, is broad. Every discipline whose claims are tested against experience qualifies as a science. Thus, sociology, anthropology, history, and geography pass muster, for evidence bears on the truth of their claims. But metaphysics, aesthetics, ethics, and religion do not. No experience could count as evidence that Rembrandt was a better painter than Rubens, that lying and intentionally misleading are morally on a par, that the spirit of God pervades the world, that behind the world of appearances there exists a world of things in themselves. Such claims, the positivists contended, are strictly nonsense since they do not satisfy the criterion of cognitive significance.[3]

The positivists were not naive. They recognized that language is often imperspicuous. But, they maintained, imperspicuousness is not nonsense, for many obscure terms can be analyzed into constructions with determinate logical forms and verification conditions. An obscure locution is cognitively significant if and only if its terms can be replaced by analytically equivalent constructions so as to yield an express tautology, an express contradiction, or a sentence that can be tested directly against experience.

---

[1] Rudolf Carnap, "The Elimination of Metaphysics through Logical Analysis of Language," in *Logical Positivism*, ed. A. J. Ayer (New York: Macmillan, 1959), 60–81.

[2] A. J. Ayer, Introduction to *Logical Positivism*, 10–17.

[3] Carnap, "Elimination of Metaphysics," 69–80.

Obviously this rough sketch glosses over a host of well-known difficulties and elegant arguments. I have said nothing about the controversies concerning what sorts of sentences admit of direct confirmation or disconfirmation. Nor have I discussed the various attempts to relax the initial demand for conclusive verification enough to preserve the cognitive significance of natural science without inadvertently readmitting Hegelian metaphysics into the cognitive realm. I do not intend to delve into such matters. My purpose in reviewing the criterion of cognitive significance is to highlight its dependence on two other positivist assumptions: the assumption that locutions admit of analytic equivalents and the assumption that individual sentences, taken in isolation, can be confirmed or disconfirmed. If these assumptions are sound, the criterion is reasonable. The line between sentences whose truth values can and sentences whose truth values cannot be established is well worth drawing. And it is plausible that sentences whose truth value cannot be established are cognitively inert for, it appears, nothing justifies preferring them over their negations. Within the positivist framework then, it is reasonable—or at least not unreasonable—to privilege science. But the assumptions in question are far from secure. They are the two dogmas of empiricism.

Because the positivists placed a premium on clarity and rigor, it was feasible to pose searching questions about the concepts they relied on and the positions they endorsed. In a cluster of revolutionary papers, W. V. Quine, Nelson Goodman, and Morton White challenged fundamental positivist assumptions.[4] Quine, Goodman, and White argued that the analytic/synthetic distinction is untenable. If they are right, philosophical theories, arguments, criteria, and concepts that rely on the distinction are in jeopardy.

They began their tradition-shattering papers by articulating and investigating a number of initially plausible explications of

[4] W. V. Quine, "Two Dogmas of Empiricism," in *From a Logical Point of View* (New York: Harper Torchbooks, 1961), 20–46; Nelson Goodman, "On Likeness of Meaning," in *Problems and Projects* (Indianapolis: Hackett, 1972), 221–230; Morton White, "The Analytic and the Synthetic: An Untenable Dualism," in *Semantics and the Philosophy of Language*, ed. Leonard Linsky (Urbana: University of Illinois Press, 1972), 272–286.

analyticity and kindred notions, and demonstrating that they are inadequate. By the time they were done, a wide variety of candidates had been canvassed and found wanting. Quine went on to sketch a picture of language that—if correct—shows why failure was to be expected. Language, he contended, is holistic. The significance of a word or sentence derives from its function in the language as a whole. In such a picture, there is no room for analyticity. No connection between words is mandatory, so no sentences are "true by definition". Nor is there a direct link between words and the world. So experience cannot conclusively verify or falsify an individual sentence. Any of many revisions might be made to accommodate recalcitrant findings. Theories get confirmed or disconfirmed, Quine maintained. Individual sentences do not.

The critique developed by Quine, Goodman, and White is no mere palace coup. For they do not propose to supplant the analytic/synthetic distinction with a different rigid dichotomy. They are well aware that their arguments tell equally against kindred dualisms—necessary/contingent, a priori/a posteriori, scheme/content, and so on. Nor are these distinctions peculiar to logical positivism. Being descendants of the Humean distinction between relations of ideas and matters of fact and of the Leibnizian distinction between truths of reason and truths of fact, such dualisms are endemic to traditional philosophy. The arguments against the dualisms thus call into question longstanding assumptions about philosophy's nature, goals, and prospects.

The denial of the dualisms was not widely and warmly received. Early on, Grice and Strawson argued that the failure of a handful of efforts to explicate a concept hardly demonstrates that there is nothing to be explicated.[5] If we consult our linguistic intuitions, they maintained, we find that we are disposed to draw the disputed distinction. As a matter of course we treat "bachelor" and "unmarried man" as equivalent in meaning, "creature with a heart" and "creature with kidneys" as equivalent as a matter of empirical fact. Without the analytic/synthetic

---

[5] H. P. Grice and P. F. Strawson, "In Defense of a Dogma," *Philosophical Review* 65 (1956): 141–158.

distinction, they contended, these dispositions would be unfounded.

Things are not quite so bad as Strawson and Grice maintained. Goodman has shown that the denial of a rigid distinction between analytic and synthetic sentences leaves room for viable conceptions of likeness and difference of meaning. Coextensive terms often belong to non-coextensive compounds. For example, although the terms "unicorn" and "centaur" are coextensive, their compounds—"unicorn-picture" and "centaur-picture"—are not. Goodman contends that coextensive terms are alike in meaning to the extent that a relevant range of parallel compounds are coextensive.[6] "Unicorn" and "centaur" then differ in meaning because the extensions of terms like "unicorn-picture" and "centaur-picture" typically diverge. "Creature with a heart" and "creature with kidneys" differ in meaning because creature-with-a-heart-descriptions usually are not creature-with-kidneys-descriptions. And "bachelor" and "unmarried man" are similar in meaning because bachelor-descriptions ordinarily are unmarried-man descriptions.[7] Likeness of meaning, as Goodman characterizes it, is clearly a matter of degree. And it is contextually circumscribed. Terms that qualify as alike in meaning in one context may qualify as different in meaning in another, for different compounds may be relevant. Likeness of meaning is thus a pragmatic matter. It depends on how symbols are used and how their uses are described. It is determinate enough to do justice to our linguistic intuitions. But it is too flexible to underwrite a rigid demarcation criterion separating cognitively significant from cognitively insignificant locutions.

---

[6] Parallel compounds are terms that result from appending exactly the same words to two terms. For example, "vixen-description" and "female-fox-description" are parallel compounds that result from appending "description" to the terms "vixen" and "female fox".

[7] Goodman, "On Likeness of Meaning." This characterization is disappointing in that it provides only an intralinguistic criterion of likeness of meaning. Using mention-selection, a device introduced by Israel Scheffler, Wolfgang Heydrich recently contrived a related criterion that applies across languages. See Israel Scheffler, *Beyond the Letter* (London: Routledge and Kegan Paul, 1979), 31–36, and Wolfgang Heydrich, "A Reconception of Meaning," *Synthese* 95 (1993): 77–94.

Our linguistic intuitions can be vindicated. But the implications for philosophy remain dire. Metaphysics totters. Without a distinction between necessity and contingency, we cannot differentiate between essences and accidents. We cannot assign context-neutral, determinate truth conditions to counterfactual conditionals, distinguish causal laws from accidental correlations, discriminate between natural and artificial kinds. There is no hope of a solution to the time-honored problem of personal identity—not because we are somehow inscrutable to ourselves, but because it makes no sense to suppose that persons (or anything else) possess an essence that makes them what they are. Epistemology too is in disarray. The project of defining knowledge is abortive, since the notion of a real definition—of knowledge or of anything else—is untenable. With the repudiation of natural kinds, the problem of induction must be reconceived. The future is bound to be like the past in one way or another. But that gives us no idea how we ought to extrapolate from the data at hand. The prospects for even a modest empiricism are thrown into doubt. If it makes no sense to say that the world impresses itself immediately and directly on the senses, it is not obvious what it is for a theory to be supported by perceptual evidence. I could go on, but by now the point should be clear. The denial of the dualisms cuts to the heart of philosophy.

This leads philosophers like David Lewis to urge that the dualisms be retained.[8] They recognize how thoroughly the disputed concepts permeate philosophy's traditional problematic, how much their repudiation would require us to give up. Since the arguments adduced by Quine, Goodman, and White do not amount to entailments, Lewis advocates retaining the dichotomies and the conception of philosophy that analyticity, necessity, and a priority underwrite.

Other philosophers—I am one of them—consider the arguments against the dualisms compelling. The history of unremitting failures to discover an adequate explication of analyticity and

---

[8] David Lewis, "Radical Interpretation," in *Philosophical Papers* (Oxford: Oxford University Press, 1983), 1:118.

a powerful philosophy of language that explains why none is to be had reinforces my view. We face a quandary. It is clear that without the dualisms there is no first philosophy. Philosophy is neither independent of nor prior to other modes of theorizing. There is no Archimedean point. Traditional philosophical problems and methods are typically framed in terms that presuppose the dualisms. What can philosophy do without them?

Opinions diverge. Physicalists like Quine answer, "Science". Nihilists like Richard Rorty respond, "Nothing". Constructivists like Goodman reply, "It depends on our interests and ingenuity". I belong to the constructivist camp.

Quine concludes that science alone survives the demise of the dualisms.[9] It generates and validates its own standards and supplies evidence of their satisfaction. If he is right, we need not look to epistemology to justify the empiricist stance. Physiology, psychology, and evolutionary biology provide the backing we need. Physiology and psychology explain how human organisms respond to events that impinge on their sense organs. Evolutionary biology affords reason to think that responding as we do has survival value. Were human responses uncoordinated with the way the world is, our ancestors would presumably not have survived long enough to reproduce. Our existence thus affords evidence of the reliability of our responses, hence of the acceptability of their deliverances.[10]

The entities science speaks of are, Quine contends, posits. Electrons, genes, brick houses, and black holes are not reducible to events that impinge on our sensory surfaces. But the posits of science are not idle speculations. They are justified by their place in confirmed theories. Scientific posits figure in theories that predict that certain sorts of events will be observed. Often such predictions are borne out. The posits then do their job. They give us reason to believe that physical objects, forces, and fields exist.[11]

[9] Quine, "Two Dogmas," 46.
[10] W. V. Quine, "On the Nature of Moral Values," in *Theories and Things* (Cambridge: Harvard University Press, 1981), 54–72.
[11] W. V. Quine, "Things and Their Place in Theories," in *Theories and Things*, 1–23.

To be sure, multiple systems of posits may serve equally well. Theory is inevitably underdetermined by evidence. So the Quinean cannot contend that science—even ideal science carried to the limits of inquiry—reveals the unique way the world is. But, Quine believes, all admissible alternatives are scientific. Wave theory and particle theory involve different posits. Both are acceptable because they belong to, and are sanctioned by, physics.

Quine makes an excellent case for the claim that science survives the demise of the dualisms. But he says remarkably little to justify his contention that science is the sole survivor. Part of the reason is surely that he shares the positivists' generous conception of the scope of science. An enormous range of scholarship falls under the heading of science, but not everything. Quine argues that intentional idioms—"believes that", "hopes that", and the like—have no place in science, since co-extensive terms are not always intersubstitutable in intentional contexts.[12] Mentalistic psychology and those works of history, sociology, anthropology, and political science that ascribe beliefs, desires, and preferences are then unscientific. So are ethics and aesthetics. Behaviorist psychology can discover what people value, and how their values are modified by conditioning, but it cannot discover what is valuable. Nor can any other science, for science reveals what is the case, not what ought to be the case.[13]

These arguments are right as far as they go, but they do not go as far as Quine thinks. To demonstrate that a discipline is not a science is not to demonstrate that it is noncognitive, unless one has already shown that only science is cognitive. Quine has not done that, or even tried to. He has, I think, simply retained the positivist contention that outside of science there is only nonsense.

Clarity, rigor, and a basis for intersubjective agreement are hallmarks of science. Perhaps, like the positivists, Quine believes they are peculiar to science. Davidson evidently believes they are

---

[12] W. V. Quine, *Word and Object* (Cambridge: MIT Press, 1960), 191–233.

[13] Quine, "On the Nature of Moral Values," 55–66.

peculiar to literal, descriptive language. Building on Quine's brilliant discussion of radical translation, Davidson argues that interpretation is a matter of mapping sentences the speaker holds true onto sentences the interpreter holds true, using a Tarski-style truth definition to provide the mapping. A few truths may be mapped onto falsehoods, but not many. For we could not understand another person, Davidson contends, unless we were largely in agreement with him.[14]

Such a method, he believes, will not yield interpretations of metaphors, since metaphors resist regimentation. He therefore concludes that terms used metaphorically have no meaning beyond their literal meaning.[15] They are false but suggestive. "Irate aardvark" denotes nothing but literally irate aardvarks. So to call a cantankerous bureaucrat an irate aardvark is to contend—falsely—that he belongs to the class of angry anteaters. Why one would make such an allegation and what others make of it are left to psychology to explain.

You can't know what isn't true. So it follows from Davidson's theory that because metaphors are almost always false,[16] they are not repositories of knowledge. We do not know that the Watergate cover-up was a cancer on the Nixon presidency, because literally it was not. Nor do we know that total science is a field of force whose boundary conditions are experience,[17] because literally it is not. It is not only isolated sentences that are discredited. Because metaphors figure prominently in literature and the other arts as well as in religion, those areas of human life are excluded from the cognitive realm. The question, Davidson believes, is not how artistic and religious symbols convey understanding, but why people falsely believe that they do. This is a question not for semantics or epistemology, but for psychology. Again, only science survives. And, although this is rarely ac-

[14] Donald Davidson, "Belief and the Basis of Meaning," in *Inquiries into Truth and Interpretation* (Oxford: Oxford University Press, 1984), 141–154.
[15] Donald Davidson, "What Metaphors Mean," in *Inquiries into Truth and Interpretation*, 245–264.
[16] The exceptions are metaphors like "No man is an island" that are trivially true under their literal interpretations.
[17] Quine, "Two Dogmas," 42.

knowledged, not all science. For actual science, like other disciplines, is riddled with metaphors.[18]

Richard Rorty agrees that science satisfies the standards it sets for itself. But he does not consider this a reason to privilege science. Literature, art criticism, cultural criticism, and political debate likewise satisfy their self-imposed standards. Nor does he think that the values of rigor, clarity, and intersubjective agreement afford grounds for favoring science.[19] We have no reason to prefer those values to flexibility, ambiguity, and creative disagreement. This is not because scientific values are in any way defective. The point is rather that without an Archimedean standpoint there is no perspective from which to assess values. The traditional philosophical projects of legitimating inquiries and validating standards neither can nor need be done. Each mode of discourse stands on its own. It succeeds if it satisfies the standards it sets for itself, whatever those standards may be.[20] The lesson Quine derives from the demise of the dualisms and the repudiation of first philosophy is that philosophy is continuous with natural science. The lesson Rorty derives is that philosophy is intellectually bankrupt. It cannot and need not shore up cognition.

Without the analytic/synthetic distinction, the verificationist criterion of cognitive significance is unfounded. If no sentence is independently verifiable, it is no criticism of "Eating people is wrong" to concede that it does not admit of independent verification. "Uranium is radioactive" does not admit of independent verification either. Nor does it follow, as Rorty thinks, that "Eating people is wrong" and "Uranium is radioactive" are just moves in a language game whose acceptability rests on nothing more than the approval of our peers.[21] We want to say—at least I

[18] See Catherine Z. Elgin, *Considered Judgment* (Princeton: Princeton University Press, 1997), chapter 6.

[19] Richard Rorty, Introduction to *Consequences of Pragmatism* (Minneapolis: University of Minnesota Press, 1982), lxi.

[20] Richard Rorty, "Method, Social Science and Social Hope," in *Consequences of Pragmatism*, 191–210.

[21] Richard Rorty, "Pragmatism, Relativism, and Irrationalism," in *Consequences of Pragmatism*, 165, and *Philosophy and the Mirror of Nature* (Princeton: Princeton University Press, 1979), 385.

want to say—that we have good reason to accept both, that in so doing we are not just parroting popular opinion, and that we do not need to adopt an (unrealizable) God's-eye perspective to see this.

If we reject the dualisms, our standards and criteria cannot be grounded in philosophical absolutes. There are none. If we reject scientism, we cannot blindly endorse the standards and criteria that natural science employs. If we reject nihilism, we cannot identify the cognitively acceptable with whatever our peers let us get away with. For that is too arbitrary—an accident of the time and place in which we happen to find ourselves. Evidently we cannot discover the standards and criteria by which our theories and other symbols are to be judged. They are not to be found. The situation is not hopeless, though, for standards and criteria can be contrived. Such is the constructivist response.

Like other branches of analytic philosophy, constructivism demands logical rigor and focuses on the ways symbols function. It agrees that symbols confront the tribunal of experience as a corporate body. But unlike Quine and his cohorts, constructivism recognizes that there are a variety of corporate structures, capable of producing different sorts of cognitive goods. It recognizes, moreover, that the corporations in questions have interlocking directorates—that, for example, even in science, fact and value intertwine. Because science is a collaborative enterprise, the moral values of truth-telling and trust are integral to it.[22] Because features like simplicity and elegance are scientific desiderata, aesthetic considerations figure in theory choice and assessment as well. We could not, I suggest, comprehend the ways science advances understanding, if we did not recognize the roles that moral and aesthetic factors play.

But we should not assume that such factors function cognitively only when they contribute to science. Goodman shows that aesthetic symbols, functioning as such, advance understanding through the arts. He constructs a taxonomy of symbol systems and shows how the syntactic and semantic structures of verbal, pictorial, and notational systems give rise to different

---

[22] See Catherine Z. Elgin, *Considered Judgment*, chapter 7.

powers and limitations. He shows, for example, that inter-subjective accord and precision trade off. The more refined a system's categories, the more difficult it is to tell which is instantiated. The arts tend to favor systems that allow for maximal precision, the sciences, systems that promise intersubjective agreement. Neither, Goodman maintains, invalidates the other.[23]

Systems of classification, Goodman believes, are human contrivances, designed to further diverse ends. They need not—indeed cannot—reflect the antecedent order of nature, for there is no such order to reflect. Rather, their rightness is determined by their effectiveness for the purposes for which they are used. The recognition that there is no fixed order of things frees constructivists to create new orders that reveal likenesses and differences that familiar modes of organization obscure. Metaphor, for example, is a device that enables us to redeploy a category scheme that literally characterizes one domain to effect a reorganization of another. To call the Watergate conspiracy a cancer on the presidency is to import into the political realm a category that literally characterizes a potentially fatal disease. The metaphor captures the conspiracy's insidious spread and grave implications. It affords an understanding of political events that no literal label captures. Metaphorical characterizations, Goodman urges, are no less determinate and no less informative than literal characterizations. They simply draw the lines in different places. There are, to be sure, no rules for the interpretation of metaphors. But there are no rules for the interpretation of literal symbols, either.[24]

Goodman recognizes a multiplicity of referential functions and devices, and shows how they equip works of art to foster understanding. Our encounters with art, he insists, enable us to see things differently. They call into question complacent assumptions, introduce new ways of ordering a domain, and provide evidence of the value of new world orders. They provide conceptual and perceptual resources that allow us to recognize features and patterns we would otherwise overlook.

[23] Nelson Goodman, *Languages of Art* (Indianapolis: Hackett, 1968).
[24] Goodman, *Languages of Art*, 68–85; "Metaphor as Moonlighting," in *Of Mind and Other Matters* (Cambridge: Harvard University Press, 1984), 71–77.

Religious symbols, Israel Scheffler argues, do too. One of the functions of ritual, he maintains, is to serve as a vehicle for cross-reference, connecting one performance with others, coalescing performers remote in time and space into an enduring community. Because each performance of the ritual refers to all the others, the ritual is a source of self-understanding. It enables its performers to see themselves as part of a religious tradition.[25]

I take it that a major consequence of the repudiation of the dualisms is that there are no natural kinds. Every two objects are alike in some respects and different in others. The world does not privilege any particular likenesses over the rest. What similarities and differences we recognize are determined by the classificatory schemes we devise, schemes which are keyed to our purposes—to the questions we want to answer, the problems we seek to solve, the constraints we want to respect, and the ones we are willing to relax. Rightness of categories is not a matter of carving nature at the joints, but of serving our purposes. And a classification that is right for one set of purposes may be wrong for another. "Grue", for example, is wrong for induction, but right for uncovering hidden assumptions about induction. Nor is there any reason to believe that a single system of categories will best serve a given purpose. Several are apt to be equally good. Thus, a demographic study might be differently but equally well served by classifications that group subjects according to income, address, or level of education. But it does not follow that every mode of classification is as good as every other. To assess a classificatory scheme, we need to consider what we want it for, what ends it is supposed to serve. The evaluation of classificatory schemes and the systems of thought they figure in then contains an ineliminably pragmatic moment.

Nonstandard modes of classification often serve our ends admirably. Classifications that cut across disciplinary boundaries illuminate functions common to symbols in the sciences and the

---

[25] Israel Scheffler, "Symbolic Aspects of Ritual II," in *Inquiries* (Indianapolis: Hackett, 1986), 64–67.

arts, and disclose continuities between ethical and scientific reasoning. Although the papers in this book belong to seemingly disparate branches of philosophy, they deploy a common set of strategies, such as the introduction of novel groupings of objects and symbols, and the recognition of multiple modes of reference. They suggest that the division of the faculties is itself an artifact of our modes of classification, and show something of the utility of reconfiguring the cognitive realm.

Constructivism is often disparaged on the ground that it is counterintuitive. Perhaps it is. But the failure of G. E. Moore's defense of common sense[26] strongly suggests that any contender for an acceptable philosophy will have counterintuitive components. It is not obvious that the position I advocate is any stranger than, say, realism about propositions or possible worlds.[27] In any case, the brand of constructivism I favor has cognitive virtues that make it an attractive alternative to currently popular positions.

One is ontological parsimony. Constructivism admits into its ontology only what is actual. It does not recognize the reality of merely possible objects or worlds. But because it does not privilege any single methodology or scheme of classification, it has the resources to construct an enormous variety of world orders, and show that new combinations of actual objects can perform roles for which possibles are often invoked.

Still, it is charged, such austerity leaves us unable to handle a host of problems—counterfactual conditionals, ascriptions of propositional attitude, fictional discourse, and so on. This charge is false. For among the things that are actual are symbols—actual utterances, inscriptions, pictures, and examples. These can be classified in ways that supply the resources we need. We can, for example, collect together a variety of pictures, descriptions, and portrayals and classify them as, say, Santa-Claus-representations,

[26] G. E. Moore, "A Defence of Common Sense," in *G. E. Moore, Selected Writings*, ed. Thomas Baldwin (London: Routledge, 1993), 106–133.
[27] For an excellent discussion of the implausibilites on all sides of the debate, see Mary Kate McGowan, "Realism or Non-Realism: Undecidable in Theory, Decidable in Practice" (Ph.D. diss., Princeton University, 1996).

or Cassandra-representations. We define the fictive characters Santa Claus and Cassandra in terms of such collections of representations. We do the same thing in a hypothetical vein. We group together a variety of descriptions under a label like "phlogiston-description" or "positron-description" or "multiple-personality-description" and investigate whether any actual thing answers to them. We need not augment our ontology with fictional or hypothetical entities. To differentiate Santa Claus from Cassandra, phlogiston from pixie dust, we can simply chart differences in the uses to which different null terms are put.

That utterances, inscriptions, pictures, and dramatic portrayals exist is beyond dispute. That they can be classified as Cassandra-representations, phlogiston-descriptions, and the like is equally untendentious. So constructivism does not extend the realm of the actual beyond the borders its realist and physicalist rivals recognize. But whereas realists like David Lewis and physicalists like Quine consider such classifications inert, we maintain that they are functional. They equip us to interpret fictional, hypothetical, and counterfactual discourse, as well as ascriptions of propositional attitude.

But, one might wonder, *how* do we recognize Cassandra-representations when we encounter them, if there is no Cassandra to test those descriptions against? We do so in the same way that we recognize instances of other one-place predicates like "constitutional democracy" or "criminal lawyer". We extrapolate from cases we already consider clear, drawing on whatever contextual cues we can find. There are no rules, but there are skills we acquire as we learn to interpret symbols of various kinds.

Individuation depends on schematization. In structuring a domain, we mark out individuals and kinds. We thereby determine what counts as the same thing, what counts as a different thing of the same kind, what counts as a different kind of thing. Caterpillars and butterflies might qualify as the same sort of thing under one schematization and as different sorts of things under another. A black hole might be classified as a burnt-out star under one schematization and as the residue of such a star under another. So whether a black hole is a star and whether a butterfly and a caterpillar are the same sort of thing turns on which system of categories is in effect.[28]

Relative to a system, it is determinate what entities and kinds there are. But absolutely, independent of all systems, such matters are indeterminate. For categorization is integral to individuation. It thus makes no sense to insist that regardless of the constructions we put on it, there exists but a single world underneath. Nor does it make sense to insist that there are many worlds—one answering to each adequate construction. For independent of all constructions, there is no basis for distinguishing one from many.[29] Still, it often does make sense to say that the objects recognized by one system are also recognized, but are reorganized, by another. In such cases, the criteria of individuation are shared by the two systems. That both the Library of Congress cataloguing system and the Dewey Decimal System are ontologically committed to books is then metaphysically unproblematic. Indeed, one of the strengths of constructivism is its recognition of the cognitive utility of reorganization. Advancement of understanding often results from reconfiguring a domain, reorienting ourselves to it, putting a different construction on familiar phenomena.

Constructivism recognizes neither a beginning nor an end of inquiry. There is no pure observational given—no unique way the world affects us, independent of and prior to all conceptualization. Every experience, thought, observation, and idea involves conceptualization—the imposition of a scheme of categories that structures the domain, highlighting some aspects, obscuring others. What we observe depends on what we look at, what we look for, and what we overlook. And these in turn

[28] See Catherine Z. Elgin, *With Reference to Reference* (Indianapolis: Hackett, 1983), 37–42.

[29] This is not widely recognized. On a standard reading, Nelson Goodman and Israel Scheffler disagree about whether, underlying diverse world versions, there is one world or many. See Goodman, *Ways of Worldmaking* (Indianapolis: Hackett, 1978), and "On Starmaking," in *Of Mind and Other Matters*, 40–42; and Scheffler, "Science and Reality," in *Inquiries*, 82–85, and "The Wonderful Worlds of Goodman ," in *Inquiries*, 271–278.

depend on the conceptual resources and expectations we bring to our looking. We test our putative insights against the standards of acceptability we endorse, and test those standards against the judgments we are inclined to accept. We emend, extend, elaborate, and refine to bring them into accord. Our goal is a system of considered judgments in reflective equilibrium. Every verdict is provisional, for currently accepted positions can be called into question by further findings. Nothing is held true come what may. But nothing is entirely arbitrary, either. For the considerations we endorse are reasonable in light of one another, and in light of what we were already inclined to accept. They are then acceptable in the epistemic circumstances.[30]

Cognition, Carnap contended, involves the construction of a conceptual framework. Internal questions, he maintained, pertain to the entities the framework recognizes; external questions pertain to the framework itself. Given a framework that recognizes animals, for example, it is an internal question whether there are herbivorous skinks. It is an external question whether animals are spatiotemporally continuous. Internal questions, Carnap believed, are factual. External questions are practical. They concern "the choice whether or not to accept and use the forms of expression for the framework in question".[31] The constructivist stance I favor is akin to Carnap's position. Like Carnap, I hold that we make the frameworks that fix the facts. But Carnap contends that the decision to accept a framework is "not of a cognitive nature".[32] I disagree: for the same sorts of considerations bear on the choice of a framework as bear on the judgments of fact internal to it. And opting for an infelicitous or inappropriate framework—for example, one that uses "grue" rather than "green" for induction—may be as cognitively disastrous as making an erroneous judgment of fact within an appropriate framework. Rightness, whether of a framework or of a statement or other symbol within a framework, is, Goodman and I have urged, a matter of fitting and working—the fitting together of the various components of a

---

[30] For a more detailed argument, see Elgin, *Considered Judgment*, 103–111.
[31] Rudolf Carnap, "Empiricism, Semantics and Ontology," in *Semantics and the Philosophy of Language*, 211.
[32] Ibid.

system of thought and the working of that system and its several components to further the ends in view.[33]

The demise of the dualisms is widely believed to constitute a massive loss for philosophy. Rorty considers it fatal to the philosophical enterprise. I contend that there are compensating gains. Once we are free of the conceptual stranglehold the traditional dualisms held us in, we can contrive a variety of frameworks, tailor made to suit our evolving interests and ends. The results are neither absolute nor arbitrary: not absolute, because they are conditional on actual, optional interests and objectives; not arbitrary, because the interests and objectives they are conditional on are not mere whims, and are themselves subject to assessment.

Between the Absolute and the Arbitrary brings together a series of papers that mark out a constructivist response to the denial of the dualisms. The papers cut a path through diverse branches of philosophy—aesthetics, philosophy of science, philosophy of language—and show how similar problems arise in each. They highlight the ineliminability of values from the realm of facts, the dependence of facts on category schemes, and the ways human interests, practices, and goals influence the categories we contrive. Individually, the papers contribute to ongoing debates in their respective fields. Collectively, the papers constitute a sustained critique of an entrenched conception of the resources available to philosophy, and argue for a constructive-relativist alternative to it.

Chapter 1 addresses the widespread conviction that natural science is agent-neutral, that it seeks to discover the mind-independent truths. Hilary Putnam purports to prove that success is practically guaranteed.[34] If he is right, nearly every ideal theory is true; and it matters little what ideals are espoused. David Lewis blocks Putnam's proof by arguing that most orders of things cannot be captured by truths.[35] Only some properties, Lewis maintains, are capable of figuring in truths. So an otherwise ideal

---

[33] Nelson Goodman and Catherine Z. Elgin, *Reconceptions* (Indianapolis: Hackett, 1988), 158.

[34] Hilary Putnam, "Realism and Reason," in *Meaning and the Moral Sciences* (London: Routledge and Kegan Paul, 1978), 125–126.

[35] David Lewis, "New Work for a Theory of Universals," *Australasian Journal of Philosophy* 61 (1983): 371; see also his "Putnam's Paradox," *Australasian Journal of Philosophy* 62 (1984): 221–236.

theory that treated of the wrong sort of properties could not be true. I argue that Lewis's absolutist account does not square with scientific practice. Scientists care about the cognitive utility of the properties they countenance; they are indifferent to metaphysical pedigree. To avoid trivializing science, I suggest, we should recognize that although science seeks truth, it does not seek every truth. It wants to discover the number of fundamental physical forces in nature, not the number of lamp posts in Harvard Yard. Science seeks only those truths that realize its cognitive values and promote its cognitive ends. Identifying and justifying suitable values and ends is a job for philosophy of science.

There is no reason to think that all sciences have or ought to have the same objectives. This, Chapter 2 argues, makes the reduction of one science to another trickier than philosophers are apt to suppose. Since science is always *in medias res*, the real question is not whether ideally acceptable theories suitably align, but whether the alignment of nonideal theories affords sufficient reason to commit disparate disciplines henceforth to develop in tandem. To decide that requires looking beyond the canonical statements of current theories to the interests, values, and perspectives of the sciences they belong to. These may diverge. If so, reduction can compromise disciplinary autonomy, requiring the special sciences to sacrifice methods, goals, and orientations that serve their cognitive purposes, but not those of the more comprehensive science that seeks to swallow them up. Even when one theory maps nicely onto another, reduction may be ill-advised.

Nor are all cognitive ends scientific. If Nelson Goodman is right, the arts function cognitively as well.[36] Art and science differ in the symbol systems they use and the values and priorities they espouse. Art, for example, allows for infinite precision, but sacrifices agreement to achieve its goal. Science sets a lower bound on significance, sacrificing precision for intersubjective accord. Neither invalidates the other. Both yield valuable insights. Chapter 3 elaborates Goodman's contributions to aesthetics and shows how the arts, like the sciences, advance understanding.

Neither does so just by disclosing facts, for isolated facts are not

[36] Goodman, *Languages of Art*.

enough. Some facts are contextually inert even when they seem clearly germane to the subject at hand. Chapter 4 argues that neither the plain facts revealed by introspection nor the more recondite facts of cognitive science solve the problem of radical translation. Even if we accord such facts the status of evidence, translation and interpretation remain indeterminate. The conviction that enough information about the speaker's psychology would yield a unique interpretation of her words turns out to be false. Beyond a certain point, psychological facts do not matter.

Facts do not speak for themselves. We have to determine what to make of them. And what we make of them may affect what we do about them. This is evident in discussions about identity through change. The problem is of more than theoretical interest to professional restorers. When restorers attempt to repair a chipped painting, for example, they apply pigment to the canvas to fill in the gaps. Do they thereby destroy the work? If the restorer's mark becomes part of the painting, it seems that they do. What was once an original Rembrandt is now a pastiche. But, Chapter 5 argues, we need not construe the restorer's mark as part of the painting. Whether restoration invariably destroys the work it seeks to repair, hence whether restoration is an aesthetically permissible act of conservation, turns on a question of interpretation. Facts determine what changes an object undergoes. They do not determine whether, having undergone those changes, it remains the same thing.

In matters of interpretation, contextual considerations loom large. The facts of psychology and neurology do not suffice to decide whether, for example, a critic believes that inpainting is invariably destructive. For they do not determine how broadly or narrowly to construe the criteria for believing such a thing. Tradition has it that our alternatives are just two: either an ascription of propositional attitude is transparent, or it is opaque. If it is transparent, every substitution of coextensive terms is permissible; if opaque, practically none is. Often neither construal is satisfactory. Chapter 6 presents an intermediate position—one that attends to both the medium of reference and the objects of reference. Extending the metaphor, I suggest that ascriptions of belief are typically translucent. They allow some, but not all, substitutions of coextensive terms.

Like belief ascriptions, metaphors are not transparent. Whatever cognitive scientists are getting at in saying, "The mind is a computer", they are not claiming that our minds are made of silicon chips. That's obvious. But to maintain that metaphors resist literal paraphrase, or that metaphorical truth is not preserved through substitution of literally coextensive terms, is not to concede that metaphors defy interpretation. Chapter 7 explicates a variety of semantic devices and shows how they function in the interpretation of metaphor. Without providing recipes, they disclose ways metaphors draw on particular aspects of context and background for their significance.

Such devices are useful in explaining how literal signs function as well. But they may look unduly complicated. C. S. Peirce paints a seemingly simpler picture when he divides signs into icons, indices, and symbols.[37] Icons and indices are supposed to be natural signs that refer to their objects directly. Symbols alone are arbitrary conventions. Chapter 8 takes issue with this account. Something is an icon, index, or symbol, I suggest, only when it functions as such. My explication reveals that rather than referring immediately and directly to their objects, icons and indices involve mediate and complex reference. They are thus highly dependent on human convention. Since interpretation is required to understand such signs, we cannot tell what they refer to just by looking. We need to know what conventions are operative, and how they operate in context.

Although such conventions admit of alternatives, they are rarely arbitrary. They belong to more or less complicated systems designed to serve our symbolic purposes. When a Monday-morning quarterback describes and illustrates a crucial play, for example, she constructs two representations of the same event. Her linguistic description draws on the full complexity of the English language, while her illustration is crude and ad hoc. Chapter 9 argues that to make sense of the parallels and diver-

---

[37] Charles S. Peirce, *Collected Papers*, ed. Charles Hartshorne and Paul Weiss (Cambridge: Harvard University Press, 1931), 1:295, and *Collected Papers*, 8:228.

gences the two representations exhibit, we need to recognize that the identity of a referring sign derives from its role as a symbol, and its role as a symbol, from its place in a symbol system. The roles a symbol can play, the distinctions it can draw, thus depend on the resources its symbol system provides. The facts turn out to be artifacts. What facts there are then depend on the symbol systems we devise.

That some facts are artifacts is hard to deny. A dollar equals one hundred cents because we have made it so. But are all facts dependent on convention? Most philosophers think not. Minimally, the facts of interest to natural science are supposed to be hard facts whose identity and character are independent of artifice. Recently, a number of philosophers have espoused a metaphysical hybrid that grafts nonrealism about social facts onto realism about natural facts. In Chapter 10 I contend that the hybrid is not viable. After arguing in Chapter 1 that the case for realism about natural facts is exceedingly hard to support, I argue in Chapter 10 that the case for the social construction of cultural categories applies to so-called natural kinds as well. Human interests and concerns figure in fixing the extension of the term "horse", just as they do in fixing the extension of the term "household". But it does not follow that we must construe all (or indeed any) of our kinds as arbitrary. Nor need we accept the limitations on intersubjective understanding that historicism leads Ian Hacking and relativism leads Bernard Williams to accept.[38] To deny that kinds are fixed in the order of nature is not to concede that anything goes.

There is, Chapter 11 argues, no sharp divide between fact and value. Far from being poles apart, they inextricably intertwine. The demarcation of facts depends on considerations of value, while assessments of value are infused with considerations of fact. Judgments of both kinds, I contend, are relative and objective. They are neither absolute nor arbitrary. There is no guarantee that they are ultimately true or permanently credible, but they

---

[38] Ian Hacking, "Making Up People," in *Reconstructing Individualism*, ed. T. C. Heller (Stanford: Stanford University Press, 1986), 222–236; Bernard Williams, *Ethics and the Limits of Philosophy* (Cambridge: Harvard University Press, 1985).

are not mere matters of opinion. They are reasonable and fallible; and they stand or fall together.

The postmodern predicament stems, Chapter 12 suggests, from differences of opinion. Such differences are of course not a recent development. But where our forebears believed that they had resolutions, we are not so sure. Nor are we clear what to make of irreconcilable differences. We might cling tenaciously to the conviction that every disagreement has exactly one resolution. Such absolutism leads to dogmatism or despair. Or we might concede that our differences are objectively irresolvable and conclude that they are therefore unreal. One answer is as good as another, we conclude; every position is arbitrary. Finally, we might recognize that there are alternative, equally good ways of understanding things. Not every difference of opinion is a disagreement, and not every disagreement has a resolution. From the fact that we differ, it does not follow that at least one of us is wrong. Nor does it follow that we are both right. Each alternative has to be considered on its merits. To say that there are several ways of being right is not to say that there is no difference between being right and being wrong. This is the most plausible option of the three.

Some interpretations, explanations, theories, and techniques, then, are tenable; others untenable. The tenable ones are those that yield an understanding of their subject matter. But there is in general no uniquely best way to understand a given subject, no way to comprehend within a single account the virtues of all the independently acceptable interpretations, explanations, theories, and techniques that apply to it.

Nor are there any guarantees. Without the abandoned absolutes, we have no way to ensure that our claims are correct. But since the theories we have managed to construct and defend reflect our considered judgments, satisfy our theoretical standards, and promote our cognitive goals, they are not mere matters of opinion. They are reasonable assessments of how things stand. Without being grounded in bedrock, they provide a relatively stable platform from which to mount further investigations. That is all we can hope for. It is all we need.

Taken together, the papers in this collection show how philosophy's bipolar disorder derives from its propensity to ignore or underrate pragmatic considerations. If nature doesn't favor one position over its rivals, we're told, nothing does. This is naive. Even if nature doesn't discriminate, *we do*. Often for good reasons. With the discovery that there is no absolute space, we can no longer baldly assert that the heliocentric theory is true and the geocentric one false. But it does not follow that each is as good as the other. We have good reason to favor the heliocentric perspective, since it admirably serves the legitimate interests of planetary astronomy. To be sure, without those interests, we might be hard pressed to decide. But that is irrelevant. For we do not claim and need not claim that heliocentrism is absolutely preferable, only that it is preferable given the interests of contemporary astronomy.

In deciding what categories we ought to employ, what criteria of identity we ought to recognize, what the facts are, and what to make of them, it is vital to consider what we want to do with our answers. We need to ask what our verdicts will be used for, what purposes they will serve, whether those purposes are worth realizing, and whether their realization will require sacrifices we are unwilling to make. There are no algorithms for answering such questions. And there is no reason to think that they have uniquely correct answers. But it does not follow that anything goes, or that one answer is as good as any other. If there are several ways of being right, there are far more ways of being wrong.

Answers on my view are neither absolute nor arbitrary: not absolute, for every acceptable framework admits of equally acceptable alternatives; not arbitrary, for every acceptable framework must answer to considered judgments about the subject, methods, and goals of theorizing. If we're careful—and lucky— they may nevertheless advance understanding.

CHAPTER 1

Unnatural Science

$\mathbf{H}$ilary Putnam's model-theoretic argument purports to ensure the truth of any ideal scientific theory—any theory, that is, that satisfies all our evidential and theoretical desiderata. David Lewis is unconvinced. Truth, he insists, is hard to come by; and our talents might not be up to the task. The risk of failure is something science cannot hope to escape. Despite the terms in which it is cast, the dispute between Lewis and Putnam is, I believe, peripheral to philosophy of science. But, I suggest, we ought not ignore it, for the fact that it is peripheral to philosophy of science is, or should be, of interest to philosophy of science and, by extension, to epistemology.

Putnam's argument is alarmingly simple. Consider an ideal (first-order) theory T_1. It is empirically adequate, answering to all observational evidence, past, present, and future. It is theoretically adequate: consistent, simple, informative, and so on (add any further theoretical virtues you like). Assume that the world consists of infinitely many things and that T_1 says as much. Then it follows from the completeness theorem that "T_1 has a model of every infinite cardinality. Pick a model M with the same cardinality as THE WORLD. Map the individuals of M one to one onto pieces of THE WORLD, and use the mapping to define relations of

M directly in the world."[1] T_1 then is true on M. Take the mapping to constitute a satisfaction relation in the sense of Tarski. Under the resulting truth definition, truth on M is truth. T_1 then is true.

The model-theoretic argument works because sets are plentiful and undiscriminating. Any collection of objects, however motley its membership, constitutes a set. So set theory has the resources to supply truthmakers for every consistent theory—not only ideal scientific theories, but also a host of other theories we have not the slightest inclination to countenance. Assuming materialism is true, for example, we secure the truth of Thales's theory by the simple expedient of assigning the class of material objects as the extension of the term "water". Under that interpretation, "Everything is water" is true.

To be sure, the class of material objects is not the intended interpretation of "water"; and truth on unintended interpretations is uninteresting. So the vindication of Thales's theory is short-lived. Does Putnam's proposed vindication of ideal theories likewise turn on the ubiquity of unintended interpretations? If so, it too is a matter of indifference.

My vindication of Thales foundered because there are constraints on the interpretation of the term "water" that my reading does not respect. But a theory is not ideal unless it incorporates whatever constraints we place on acceptable interpretations of its terms. All constraints on the acceptable interpretation of the term "water" then are built into the ideal theory of water, even if not into my construal of Thales. Every mapping of ideal theory onto the world supplies an intended interpretation. For whatever satisfies all our constraints answers to our intentions. We can, of course, introduce further constraints if we think we've been too lenient. But any additional constraints we decide to impose are just more theory. The augmented theory will have different models from its predecessor, but it too is subject to the model-theoretic argument. If it is consistent, it has a model that affords a mapping under which the theory comes out true. If Putnam is

---

[1] Hilary Putnam, "Realism and Reason," in *Meaning and the Moral Sciences* (London: Routledge and Kegan Paul, 1978), 125–126.

right, there is nothing *we can do* to evade this result. For whatever we do is just more theory. I think he is right. So does Lewis.

But, Lewis notes, this does not make the result unavoidable. Reference is a two-place relation. What one of the relata cannot do, the other may be able to. So even if *we* cannot disqualify motley collections from serving as extensions of our terms and truthmakers of our theories, it does not follow that they are qualified. For the world may disqualify them for us. Lewis contends that unless the world privileges certain schemes of organization, and precludes reference to sufficiently gruesome properties, Putnam's result is inevitable.[2] I agree. Lewis concludes that the world confers such privilege. I conclude that Putnam's result holds.

I do not see how the world could privilege properties. But there is no premium in getting into a battle of incredulous stares with David Lewis. Doubtless there are more things in heaven and earth than are dreamt of in my philosophy. Lewis's elite properties may be among them. In any case, the rest of this chapter is predicated on the assumption that I do not have to understand *how* the world confers privilege in order to investigate the implications of the hypothesis *that* it does.

One caveat: I am concerned here only with the implications for philosophy of science. Commitment to elite properties runs deep in Lewis's metaphysics, and the character and fate of empirical science are not among his central concerns. So even if my argument is entirely successful, Lewis has ample reason to continue to countenance such properties. The issue is whether those who don't buy into the rest of his metaphysical position need do so as well.

The model-theoretic argument trades on the abundance and democracy of sets. Lewis acknowledges the reality of all the sets and recognizes a property for each set. But, he maintains, some sets—indeed the overwhelming majority of sets—are so miscellaneous in their membership that they and their properties are too gruesome to refer to. Only an elite minority of properties supplies

---

[2] David Lewis, "New Work for a Theory of Universals," *Australasian Journal of Philosophy* 61 (1983): 371; see also his "Putnam's Paradox," *Australasian Journal of Philosophy* 62 (1984): 221–236.

candidates for reference. These Lewis calls perfectly natural properties. It is possible to refer to them and to properties definable in not-too-complex ways out of them, and that's all. The vast majority of properties, being plebeian, are ineligible for reference. Since truth requires reference, plebeian properties do not figure essentially in any truths.

Assuming that perfectly natural properties are not too plentiful and that the restrictions on definability are not too lax, the position Lewis advocates evades Putnam's result. Although an ideal theory is sure to have an intended model in the world, the properties the model marks out need not be eligible for reference. If they are not, the theory fails to refer. In that case, it is not true.

The real metaphysical divide comes between perfectly natural properties and all the rest. But perfectly natural properties constitute too exclusive an elite to secure all reference. We could not even refer to atoms, much less to cats or carbohydrates, if reference were restricted to perfectly natural properties. Simple set-theoretical compounds seem innocuous enough. So, Lewis contends, simple set-theoretical compounds of perfectly natural properties qualify as natural properties as well. Naturalness grades off. Some properties lie at a sufficient remove from perfectly natural properties that there is no fact of the matter as to whether they are eligible for reference. Their referential status is indeterminate. Even more remote properties are decidedly unnatural. We cannot refer to them.

This does not seem particularly objectionable. Like eligibility of bachelors, eligibility of properties wanes when entanglements become too complicated. And even if we're not sure exactly where the boundary lies, eventually it is obvious that we're on the other side.

To be fair to Lewis, then, we should not contrast "*green*" with "*grue*", as Goodman defines it. For both may designate manifestly eligible properties, or "*grue*'s" referential status may be indeterminate. (I take it that "*green*'s" eligibility is unproblematic.) Rather, we should contrast "*green*" with "*grue**", which is equivalent to "examined before 3000 and found to be green, or examined between 3000 and 3005 and found to be blue, or examined be-

tween 3005 and 3010 and found to be yellow, or, . . . ," adding
as many disjuncts as necessary to make the property clearly
ineligible.

It might seem that my concession that we reach ineligible properties by adding disjuncts makes Lewis's point for him. If it takes,
say, thirty seven disjuncts to define "grue*", and we can't wrap
our minds around such a large disjunction, doesn't it follow that
we cannot refer to *grue**?

Not obviously. You do not have to be able to define a property
to be able to recognize its instances. And if you can recognize
things as instances of it, you can refer to it. I cannot define
"green", but I readily recognize some green things. That being so,
I readily recognize some grue* things and readily recognize that
they are grue*. To the objection that I cannot tell of other things
whether they are grue*, the reply is that I cannot tell whether
those very same things are green, either. They have not been
observed. To decide on the instantiation of either predicate
requires examining the object. And once the object is examined,
its instantiation of both properties will be settled simultaneously.
If our inability to tell whether unexamined objects are green
does not prompt us to conclude that *green* is ineligible for
reference, it should not prompt us to conclude that *grue** is ineligible, either.

In any case, eligibility for reference is supposed to be independent of the psychology of would-be referrers. So the human inability to handle thirty-seven-place disjunctions cannot be what
rules out *grue**. If *grue** is ineligible, human incompetence is not
what makes it so. Eligibility, Lewis maintains, is entirely objective. It in no way turns on our having any special—or indeed
any—access to the properties that possess it.

Lewis believes that the properties we recognize and those we
have access to are natural properties. This may be overly optimis-
tic. Humans evolved in the actual world and developed capacities
to discern properties critical to mundane survival. If the actual
world is sufficiently gruesome—and we have no guarantee that it
is not—the properties we are adapted to discern may be decidedly unnatural. Consider, for example, the predicate "toxic".
Apart from the capacity to kill or sicken the organisms they affect,

its instances have little in common. In particular, their molecular structures are quite diverse. There is no particular reason to think that *toxic* is sufficiently closely related to perfectly natural properties to qualify as a natural property. Were it not for our aversion to illness and death, the class of toxicants is one we would have little reason to remark. Maybe the point generalizes. The other properties we have occasion to notice might be equally parochial. Perhaps we are adapted to discern them only because they answer to species-specific interests and needs. If so, their claim on our attention does not derive from their displaying any special uniformity apart from those interests and needs. And conceivably, simple set-theoretical combinations of perfectly natural properties mark out sets that humans have no reason to notice, since they do not especially thwart or promote our objectives. Ineligible properties might easily command our attention and epistemic allegiance more strongly than eligible ones. That is why ideal theories might be false.

Perfectly natural properties, Lewis believes, carve nature at the joints and figure in the fundamental laws of physics. Since other eligible reference classes obtain their eligibility via chains of definability from perfectly natural properties, the laws of physics—as Lewis construes them—delimit the whole truth about the actual world. Although there is much that cannot be referred to, no truth remains untold. So physics's claim to comprehensiveness is upheld. The question is whether physics as physicists do it has any reason to seek or favor perfectly natural properties and the laws they figure in. If not, physics as physicists do it is not the enterprise Lewis describes. It is not natural science.

The worry is this: the factors that distinguish eligible from ineligible referents must be independent of us; otherwise they are of no help against Putnam. But if they are genuinely independent, it is hard to see how they can do any epistemological work. Have we any reason to believe that the considerations that influence theory choice or assessment key into natural properties? Lewis thinks we have.

High on anyone's list of scientific virtues is simplicity. *Ceteris paribus*, the simplest theory compatible with the evidence is to be preferred. But it is not obvious what makes for simplicity; nor is it obvious why simplicity is something science should value.

The problem is that simplicity seems to be a function of language. We are inclined to consider

> All emeralds are green

simpler than

> All emeralds examined before *t* are grue; all other emeralds are bleen.

But both amount to the same thing. So science should consider them on a par. We may be inclined to consider

> All emeralds are green

simpler than

> All emeralds are grue

even though "green" and "grue" are alike one-place predicates and each can be defined in terms of the other. If we want to vindicate such inclinations, we need a basis for discounting simplicity of formulation in a language that takes "grue" as primitive.

In any case, if simplicity is an artifact of the language in which a theory happens to be formulated, science's reason for valuing it is hard to fathom. Why isn't simplicity like sonority or meter or rhyme, a matter of mere aesthetic interest? Science would hardly favor a theory over its rivals merely because it sounded better in French!

Lewis has the resources to answer both questions. The simplicity science seeks is not simplicity of formulation in whatever language we happen to speak, but simplicity of formulation in a language whose primitives refer to perfectly natural properties. So the ground for choosing between

> All emeralds are green

and

> All emeralds are grue

is that when theories incorporating both are translated into such a language, one is simpler than the other. A vocabulary that refers to perfectly natural properties supplies a scale on which theories can be objectively compared. Since that vocabulary divides nature at the joints, the scale it supplies is neither arbitrary nor just more theory.

This is an elegant evasion of Putnam's result, but a couple of worries remain. Currently, properties such as quark color and flavor are the prime candidates for the status of perfectly natural properties. Still, it is the actual structure of things, not the structure we, or our wildly successful scientific successors, ascribe to it that settles the issue. We do not now know what the world's perfectly natural properties are. And even if we optimistically assume that our current best candidates—quark colors and flavors and the like—are, or come very close to being perfectly natural properties, we have no idea how to construct chains of definability linking them with other entities and kinds that science countenances—polypeptides, ecological niches, spiral galaxies, and so on. Since we do not know what perfectly natural properties are or how to define other properties in terms of them, they cannot figure in our assessments of simplicity. Judgments must be made on the basis of available information.

In fact, scientists assess simplicity on the basis of whatever scheme of classification happens to be in effect. Contemporary geologists cheerfully use contemporary geological categories, undaunted by the knowledge that historical contingencies affected their development. We do not hold out for a translation of "green" and "grue" into an as yet unknown vocabulary that divides nature at the joints before we feel safe about calling "green" simpler than "grue". And I doubt that we would reverse our verdict should it turn out that the connection of "grue" to quarks is slightly less circuitous than that of "green".

The simplicity scientists are in a position to assess is simplicity relative to a scheme of organization whose evolution was fraught with contingencies. Such a scheme affords a conceptual structure within which investigators can pursue their inquiries. They can say this much in its defense: they know none better. But there is no reason to believe it affords the only such structure or the best one. At various points in the development of the discipline, choices had to be made, balances among competing factors struck. The current scheme is the product of such choices.

The simplicity of such a scheme might seem too superficial to be a scientific value. But it is the only simplicity available to working scientists. The objective simplicity Lewis recognizes may be a legitimate end of science, but it cannot be a means of achieving science's ends.

Simplicity is in any case neither the sole scientific desideratum nor the overriding one. Science has multiple desiderata—scope, informativeness, precision, and the like—that can and often do vie with simplicity and with each other. An acceptable theory strikes a balance among them. Tradeoffs have to be made. A gain in informativeness compensates for a loss of simplicity, or a gain in precision for a loss in scope. Frequently, there is ample reason to sacrifice some measure of simplicity to achieve a better theory overall. So if simplicity is the hallmark of naturalness, science often has reason to go unnatural. For the theories that display the best balance of scientific desiderata are not in general the simplest. Not that simplicity is peculiarly defective in this regard; to take some other desideratum as the mark of naturalness would avail us nothing. All scientific desiderata are susceptible to tradeoffs; none inevitably trumps.

Good science then can veer away from Lewis's natural properties. A theory that treats of unnatural properties may be as good or better than one restricted to natural ones.[3]

Suppose this occurs. Suppose, that is, that science develops a maximally general theory that answers to the evidence and is on balance as simple, informative, predictive, and precise as any of its rivals. It is in fact an ideal theory. Lewis contends that it still might be false; for it might not be cast in terms of natural properties.

[3] Bas van Fraassen, *Laws and Symmetry* (Oxford: Clarendon Press, 1989), 40–59.

Suppose it is not. Still, its terms pick out real properties—if plebeian ones. The objects it assigns to the extensions of its predicates in fact belong to those extensions. Its sentences are accurate under their intended interpretations. So things are just as the theory says they are. Its predictions are borne out. The generalizations it accords the status of laws form an integrated system that subsumes events and explains why they were to be expected. How could this be considered a scientific failure? What more do you want?

Truth, replies Lewis. An acceptable scientific theory should do all of the above while restricting itself to natural properties. For only a theory that restricts itself to natural properties has what it takes to be true.

Rumblings begin to be heard in the neighborhood of the Bastille. What is it that is supposed to qualify elite properties and disqualify plebeian properties from figuring in truths? What do elite properties have that plebeian properties lack? Well, the elite, natural properties are more closely related to perfectly natural properties than plebeian properties are. But that just pushes the problem back a step. Aristocratic lineage is impressive only if it relates its possessors to something worth being related to. What is so impressive about perfectly natural properties that we should restrict reference and truth to them and their kin?

Supervenience suggests itself here. If plebeian properties all supervene on natural or perfectly natural ones, it might be urged, their status is derivative, and their claims on our epistemic allegiance dubious. The eliteness of natural properties then consists in their being metaphysically more basic than plebeian properties. But supervenience cuts both ways. Elite properties likewise supervene on a suitable array of plebeian properties. So relative to a different basis, elite properties are derivative, and their claims equally dubious.

Nor can we contend that a theory that treats only of natural properties better satisfies our evidential and theoretical desiderata. Tradeoffs will yield multiple theories that are equally good on balance. So the process of theory construction allows for ties. Moreover there is no particular reason to think that a theory that restricts itself to natural properties will be among those tied for first place.

Our parochial concerns pretty clearly influence our choice of categories for macroscopic objects and kinds. Historical and cultural circumstances; physical capacities and limitations; actual, optional interests and priorities affect the schemes of organization we contrive. It would be rather remarkable if, for example, a taxonomy that draws the distinction between horses and zebras where we do aligned at all well with natural categories that were suitable for describing the cosmos as a whole, but were indifferent to human faculties and ends.

It is not that the distinctions we draw are somehow ad hoc. They mark real differences in the world. The problem is that there are too many real differences in the world. Not all are worthy to mark off kinds. And, I suggest, a sufficiently disinterested perspective might find no reason to draw the lines where we do. Were it not for our interest in domesticating animals, for example, the (real) differences between horses and zebras might as easily be construed as differences within a kind as differences that mark a distinction between kinds.

Nor are our microscopic categories clearly unaffected by human contingencies. Isotopes could as easily be construed as different types of atoms or as variants of the same type of atom. Given certain reasonable interests that predominated in the development of physics, it makes sense to construe them as variants. But if we bracket those interests, there may be no ground for favoring either construal over the other. Physics could, I suggest, progress equally well under either scheme.

There is no reason to think that only theories that restrict themselves to natural properties will satisfy our scientific desiderata. Two other outcomes need to be considered. A worst-case scenario for Lewis is this: science yields ideal theories. They are on balance maximally simple, informative, explanatory, and evidentially adequate. But the properties that figure in them are plebeian. The best theory that treats exclusively of natural properties is decidedly less than ideal. Should science favor the latter? Lewis, I think, has to say "yes". For the natural theory yields truths—even if less elegant, predictive, and precise truths than we'd hoped for. Whatever other virtues they may have, the unnatural theories are false. Working scientists and, I venture, most philosophers of science would disagree. It would be scientifically irresponsible to scrap an ideal theory in favor of a less than ideal one, however elite the properties of the latter.

A slightly less clear case is this: suppose that among the theories that satisfy our theoretical and evidential desiderata is one that restricts itself to natural properties. Call it a natural theory. Does that restriction supply sufficient grounds for favoring it over unnatural ideal theories? Again, I think, Lewis would say "yes". But even here, the answer is not obvious. If naturalness is not a scientific virtue, it is irrelevant to theory choice. Whether an ideal theory is a natural theory is a matter of indifference. So suppose that it is a scientific virtue. Then the naturalness of its properties is a theoretical asset of the ideal natural theory. Ideal plebeian theories, being deficient in naturalness of properties, must compensate for this deficiency by displaying other virtues; otherwise they would not be ideal. Perhaps they are more precise, or more elegant, or better predictors than the ideal natural theory. But if the theories are equally good on balance, science still has no reason to favor the natural one. It would not be scientifically irresponsible to prefer the natural theory. But it would not be irresponsible to prefer any of its plebeian rivals, either, or to have no preference. A tie is a tie.

If at some ideal end of inquiry, science arrived at a theory that provided a suitable balance of scientific desiderata, Putnam's argument shows that the theory would have an intended model in the world. The individuals it recognized would exist and would instantiate the properties and stand in the relations the theory ascribes to them. The generalities it accords the status of laws would reflect regularities that we could reasonably accept in advance of examining every instance. The system of generalizations it treats as laws would subsume events and, in light of accepted initial conditions and background assumptions, explain why those events were to be expected. The generalizations would moreover yield accurate predictions. In short, those generalizations would be both true on an intended model and lawlike. They ought, I suggest, be accorded the status of laws. Not, if Lewis is right, natural laws, since the properties they invoke do not belong to the natural elite, but laws nonetheless.

Lewis has contrived a new form of skepticism. Even if we knew that the world has the structure our science ascribes to it, we might still be wrong; for although that structure is a genuine structure, it might still be the wrong structure. The order we have discovered might not be the natural order. Indeed, the actual world might be so gruesome that natural laws, reflective of elite properties, are laws human beings have no reason to believe. Nevertheless, we would be wrong to believe the elegant, informative, accurate generalizations that ideal unnatural science discloses. For the unnaturalness of their properties debars them from truth.

Lewis's natural properties constitute a metaphysical aristocracy. They are the elite whose standing derives from the refinement of their antecedents, not their contribution to the cognitive enterprise. But science, I suggest, is a meritocracy. It accords respect to categories, methods, approaches, and instruments that prove themselves through their contributions to the advancement of understanding. It makes no difference to their current standing whether scientific categories began their careers as members of a metaphysical elite or as ad hoc, gerrymandered expedients cobbled together out of desperation. For the purposes of science, all schemes of organization that enable us to make maximally good sense of things are equally worthy, and are preferable to any scheme that at its best enables us to make less good sense than its rivals. And making good sense has to be measured by our own standards; for we have no other.

Lewis's conviction that the naturalness of properties is independent of and antecedent to scientific inquiry recalls the Euthyphro problem. In *The Euthyphro*, the question was, roughly: Is an action good because the gods like it, or do the gods like it because it is good? The counterpart is: Does science favor particular properties because they are natural, or are they natural because science favors them? Plato made it quite clear that the answer to the Euthyphro question was supposed to be that the gods like an action because it is good—because, that is, it has some independent, antecedent, good-making properties. Without commenting on the persuasiveness of Plato's argument about goodness, I suggest that the answer to our question does not run parallel. Nothing confers naturalness on properties but their contribution to successful science. Properties are natural then only because natural science favors them. Naturalness of properties is an output of successful inquiry, not an input into it.

I said earlier that the resolution of the dispute between Lewis and Putnam is of little interest to philosophy of science. But the gist of my argument so far seems to be that Putnam wins. How can that *not* be interesting? That ideal theories are bound to be true does not look like a trivial result.

The answer is that Lewis is right in thinking that Putnam makes it all too easy. Truth, it turns out, is cheap. It is a mere side-effect of ideal theories.

Of course, "too easy" is being used here in a distinctive, philosopher's sense. Actually coming up with a theory that even remotely approximates our ideals is enormously difficult; and scientists might never succeed. Still, being a philosopher, I think there remains a legitimate sense in which it is too easy.

This can be seen in the vagueness and inconstancy of my talk of scientific desiderata. In discussing what makes for an ideal theory, I rounded up the usual suspects and generously allowed you to complete the list as you please. If you like symmetry, put it on the list; otherwise leave it off. If you favor materialism, restrict your ontology to material objects; if not, impose no such restriction. It matters not at all. Almost whatever your ideals, the theory that satisfies them will be true. The only constraints that figure in Putnam's proof are consistency and size. For the proof to succeed, an ideal theory must be consistent; and it must contend, and be right in contending, that the world contains infinitely many things. But if that's all it takes to get truth, if all other cognitive ideals are optional, truth does not seem a prize much worth coveting. Contrary to popular opinion, truth is not the elusive, hard won goal of scientific inquiry. It is an all but inevitable side-effect of satisfying scientific ideals, whatever they may be. Rather than arduously, single-mindedly pursuing truth, science more or less trips over it.

Truth is cheap—far too cheap to qualify as the end of inquiry. For if it were our ultimate scientific goal, we could easily achieve it: adjust our desiderata until available theories qualify as ideal, declare victory, and go home. Even in this age of diminished expectations, aspirations, and funding, no one is quite that jaded.

For philosophy of science, the important lesson of the Putnam–Lewis dispute is that we cannot construe (mere) truth as the end of scientific inquiry. Not, as the skeptic contends, because truth is too hard to come by, but because it is too easy.

Truth is guaranteed, practically irrespective of our cognitive desiderata. So we can't justify particular desiderata by claiming that their satisfaction is especially conducive to getting truth. What can we say in their favor? Adopt one set of desiderata, and complexity is a cognitive virtue. Adopt another, and simplicity is. Both can figure in ideal theories. So what—if anything—is to be said in favor of simplicity? Is there just no disputing matters of theoretical taste? Rather, I suggest, the answer is this: in our scientific investigations, we seek an understanding of a certain kind. The value of simplicity, scope, informativeness, and the like derives from their contributing to the advancement of that sort of understanding. Their merit lies not just in their yielding truths, but in their yielding truths of a certain kind.

This opens the way to questions about what sort of understanding science, or some particular science, is or ought to be after—that is, about what desiderata it does or ought to accept. Thus, for example, Bas van Fraassen contends that in philosophy of physics at least, we should take symmetry concerns more seriously and moderate our enthusiasm for laws.[4] Jerry Fodor advocates letting each science set its own agenda rather than insisting that all draw their ideals from a common store. Desiderata for ideal physics need not be all or only the desiderata for ideal psychology.[5]

The issue is not whether the science in question will yield truths, but whether it will yield the sorts of truth it seeks. But to construe the matter this way might seem to put us back in Lewis's

[4] Ibid.
[5] Jerry Fodor, "Special Sciences," *Representations* (Cambridge: MIT Press, 1981), 127–145.

camp. Science, I am suggesting, is not just after truth. It is after the right sort of truth. Elitism looms.

The difference is this. The criteria for being the right sort of truth are constructed by the science, not supplied by the world. In developing and justifying and modifying and balancing its several desiderata, a science delineates the sort of truths it is after and the sort of understanding it seeks to provide. But there is no reason to think that all sciences will or should consider the same sort of truth the right sort. Even if physics places a premium on exceptionless laws, Fodor contends, psychology need not follow suit; for its interests and priorities may be different.

Language consists of syntax, semantics, and pragmatics. Syntax concerns form; semantics, content; and pragmatics, use. Philosophy of science takes syntax and semantics seriously. But it typically dismisses pragmatics on the grounds that utility is irrelevant to truth. My discussion suggests that this dismissal is premature. Utility may be irrelevant to whether a claim is true, but it is not necessarily irrelevant to whether its truth value is of interest to science. In constructing, balancing, modifying, and adjudicating among our desiderata, we contrive standards of cognitive utility—standards that delimit the truths that are worth seeking. One of the lessons of contemporary philosophy is that form is not detachable from content. My discussion here suggests that pragmatic considerations are not detachable either.

Having come this far, we might wonder whether all legitimate cognitive interests are reflected in science, however broadly construed. There is no particular reason to think so. Thus, Daniel Dennett argues for the independent interest of the deliverances of the intentional stance, even though he doubts that they can be captured by science.[6] And Nelson Goodman argues for the cognitive contributions of art, which he flatly denies can be accommodated by science.[7] We might even venture to suggest that philosophy yields a measure of understanding, without donning for it the ill-fitting mantle of science.

[6] Daniel Dennett, *The Intentional Stance* (Cambridge: MIT Press, 1987), 43–68.
[7] Nelson Goodman, *Languages of Art* (Indianapolis: Hackett, 1968), 262–265 and *passim*.

If in the end our justification for the values of science is conditional on scientific interests, epistemology has the task of asking what, if any, other cognitive interests and values are worth realizing. Science has no monopoly on truth. It need not have a monopoly on interesting truth either.

CHAPTER 2

Theory Reduction: A Question of Fact or a Question of Value?

Reduction of scientific theories is often treated as a linguistic issue: one theory is said to be reduced to another if the objects referred to by the former are identified with entities in the domain of the latter, and the laws of the former are derived from the laws of the latter (plus whatever connecting principles, correspondence rules, or bridge laws are needed to link the vocabularies of the two theories). The objects of the reduced theory are thus shown to be *nothing but* objects (or combinations of objects) recognized by the reducing theory, and the concepts of the reduced theory are shown to be theoretically superfluous. The resulting ontological and conceptual economies may reasonably be construed semantically, for they demonstrate that the language of science requires fewer primitive terms than had been previously supposed. Positions taken by Quine and Goodman suggest, however, that evidential and linguistic arguments are in principle too weak to secure theoretical reduction.

The failure to discover a syntactic or semantic solution to Goodman's paradox suggests that there are no formal criteria for determining which generalizations are lawlike. Let "*x* is grue" be defined as "*x* is examined before time *t* and found to be green or *x* is not so examined and is blue". There appear to be no formal arguments to demonstrate that "All emeralds are green" is lawlike, but "All emeralds are grue" is not. Why then should we accept the former rather than the latter since (*t* being some future time) both are equally compatible with the evidence? This is the new riddle of induction.

Quine argues that language must be conceived holistically. With the repudiation of the analytic/synthetic distinction we lose any objective ground for separating out favored usages as giving the meaning of a term while taking others to convey contingent beliefs about its objects. Indeterminacy of translation and ontological relativity are consequences of the holism of language. So long as dispositions to verbal behavior are preserved, there is no factual basis for preferring one translation manual to another. And we individuate objects by means of a system of linguistic devices—"plural endings, pronouns, numerals, the 'is' of identity, and its adaptations 'same' and 'other'".[1] Thus the reference of our terms and the extensions of our predicates are relative to the interpretation of those devices. There is then no saying absolutely what the terms of a language refer to or whether two languages refer to the same things.

If syntactic and semantic arguments are incapable of determining which generalizations are lawlike, and incapable of deciding in any absolute sense what the objects of a theory are, they are (to put it mildly) unlikely to be sufficient to demonstrate that the objects of one theory are identical to (some of) those of another, and that the laws of the one can be derived from the laws of the other. Accordingly, even if the result of theoretical reduction is a change in the syntax and semantics of the language of science, it should not be thought that only linguistic and evidential arguments are required to achieve this result. Or so I will argue.

The structure of my argument is this: I begin by considering Davidson's argument that anomalous monism yields a solution to the mind/body problem. Anomalous monism is the thesis that every mental event is physical, but it is not the case that mental events can be given purely physical explanations. Davidson recognizes the validity of Quine's arguments. But he claims that the

[1] W. V. Quine, "Ontological Relativity," in *Ontological Relativity and Other Essays* (New York: Columbia University Press, 1969), 32.

formal question of lawlikeness—that is, whether a generalization, if confirmed by the evidence, is to be taken as lawlike—is a question of the "fit" between the predicates in the generalization. "Nomological statements bring together predicates we know a priori are made for each other."[2] Thus on his account "All emeralds are grue" is not lawlike because "emerald" and "grue" are not suited to each other. Correspondingly, he denies that psychophysical generalizations are lawlike because mental and physical predicates are not suited to one another. He argues that the indeterminacy of translation demonstrates that psychological and psychophysical generalizations cannot be reduced to purely physical laws: There are indefinitely many mappings of them onto our physical theory and there is no fact of the matter as to which one is correct. Although he accepts the thesis that every mental event is in fact identical to some physical event, Davidson concludes on the basis of linguistic arguments that there are no laws linking events described as mental with events described as physical. In effect, he believes that the mind/body problem can be solved by the philosophy of language.

I argue that the linguistic arguments he adduces are not sufficient for his conclusion (or if they are, then they constitute a general a priori argument against theoretical reduction). Nevertheless, Davidson's account is important for at least two reasons: First, it demonstrates the viability of a token-token identity thesis. It may be the case that every object in the domain of one theory is identical with an object (or combination of objects) in the domain of another, without its being the case that the kinds recognized by the former are identical with kinds recognized by the latter, or that laws expressed in terms of the conceptual apparatus of the former are reducible to laws expressed in terms of the conceptual apparatus of the latter. Second, although his appeals to language are not themselves sufficient, the intuitions that motivate them may be based on sound reasons for blocking reduction.

[2] Donald Davidson, "Mental Events," in *Experience and Theory*, ed. Lawrence Foster and J. W. Swanson (Amherst: University of Massachusetts Press, 1970), 93. Page numbers in parentheses throughout this chapter refer to "Mental Events".

Davidson's account suggests that different disciplines may have conceptual commitments that preclude reduction even when the requisite correlations can be established.

I hope to show that the decision as to what generalizations ought to be treated as lawlike and the decision as to whether the objects of one theory ought to be identified with the objects of another are normative decisions. They are based on the interests and cognitive values of the different disciplines, and perhaps also on more general values that those disciplines are designed to serve. If the disciplines differ in the values that they recognize, theory reduction may be blocked even though the evidence supports correlations between generalizations warranted by the two theories.

My discussion focuses on the case of psychophysical reduction, both because previous discussions of token-token identity have done so, and because it is fairly easy to show that normative issues are relevant to deciding this case. I think that these issues arise for other proposed theoretical reductions as well—even for the celebrated reduction of chemistry to physics. But since the normative dimensions of our activities are typically brought to our attention by *disagreements*, if our investigation of reduction is restricted to disciplines that share cognitive values, the fact that there is a normative aspect to the problem is likely to be overlooked.

Davidson draws the distinction between the mental and the physical linguistically: A term is mental if and only if it is intentional; otherwise it is physical (84). Accordingly, an event is mental if and only if it uniquely satisfies an open sentence that employs at least one intentional term essentially. A mental event is thus the object of a propositional attitude. An event is physical if and only if it uniquely satisfies an open sentence that employs only physical terms essentially. It follows that the same event can be both mental and physical.

Davidson's argument rests not on the de facto failure of psychology to discover rigorous, exceptionless, predictive laws, but rather on the conviction that psychological and psychophysical generalizations are not fully lawlike. Determining under what circumstances a generalization is lawlike is thus of central importance for Davidson's philosophy of mind.

Generalizations that are lawlike are supported by their instances and sustain counterfactuals. Since evidential support is never strong enough to require us to recognize a generalization as lawlike, "ruling it lawlike must be *a priori*" (90). Like his criterion of the mental, Davidson's criterion of lawlikeness is linguistic. He claims that "lawlikeness is a matter of degree" (92), and argues that the degree of lawlikeness of a generalization is a function of the fit between its terms. He does not explain what it is for predicates to be "made for each other" or "suited to each other" (92). Like the lawlikeness of the generalizations that result, the compatibility of predicates is a matter of degree. What is unclear is the nature of the a priori linguistic knowledge which is supposed to determine the intimacy of the relation between predicates of a given generalization, and hence the degree of inductive support its instances can afford.

Although "mental and physical predicates are not made for one another", Davidson claims that psychophysical generalizations are more lawlike than "All emeralds are grue". We apparently know a priori that psychophysical claims are well formed and can be confirmed by evidence. Such claims in turn confirm rough psychophysical generalizations which are lawlike to the extent that they provide "good reason to expect other cases to follow suit roughly in proportion". Although they are "assumed to be only roughly true, or . . . are explicitly stated in probabilistic terms, or . . . are insulated from counterexample by generous escape clauses" (93), these generalizations (like those in Hempel's explanation sketches) can be refined to the point where they state precise, exceptionless, predictive laws. The critical issue is what form the refinement takes—specifically, in what vocabulary such refinement must be expressed.

On the one hand, there are generalizations whose positive instances give us reason to believe the generalization could be improved by adding further provisos and conditions stated in the same general vocabulary as the original generalization. Such a generalization points to the form and vocabulary of the finished law: we may say that it is a *homonomic* generalization. On the other hand there are generalizations which when instantiated may give us reason to believe that there is a precise law at work, but one that can be stated only by shifting to a different vocabulary. We may call such generalizations *heteronomic*. (94)

A homonomic generalization is one whose vocabulary is that of a comprehensive closed theory. Unless a theory is comprehensive and closed, events within its domain causally interact with events outside of its domain. Since the latter are not describable in the vocabulary of the theory, such causal sequences are not subsumable under the laws of the theory. In that case the theory is unable to explain the occurrence of every event within its domain. Davidson contends that only homonomic generalizations are fully lawlike, for heteronomic generalizations, by their very form, require replacement rather than revision to yield explicit, precise, exceptionless laws.

Davidson's argument for the anomalism of the mental rests essentially on the heteronomic character of psychophysical generalizations. This seems a bit strange, for the question one wants to ask is whether mental *events* are or are not governed by explicit, predictive, exceptionless laws. This question at least seems to be independent of questions concerning the semantic character of our descriptions and generalizations regarding such events. Davidson contends, however, that such independence is illusory. Mental events are identified as the objects of propositional attitudes. Indeed, "events are mental only as described" in the vocabulary of propositional attitudes (89).[3] Hence, if terms belonging to that vocabulary are incapable of entering into statements of genuine laws, events identified as mental are anomalous. Because of the description-relative character of the mental, any attempt at redescription is precluded. "[T]o allow the possibility of . . . laws [linking the mental and the physical] would amount to changing the subject, . . . deciding not to accept the criterion of the mental in terms of propositional attitudes" (90).

---

[3] Note that on this criterion, objectless emotions like happiness and attitudes like pessimism would not qualify as mental.

Propositional attitudes constitute a network and the content of each propositional attitude depends on its place in that network. Accordingly, a mental term derives its meaning from its place in the descriptive system representing the network of propositional attitudes. "There is no assigning beliefs to a person one by one on the basis of his verbal behavior, his choices, or other local signs no matter how plain and evident, for we make sense of particular beliefs only as they cohere with other beliefs, with preferences, with intentions, hopes, fears, expectations and the rest" (96). Davidson's point is not epistemic, but ontological—not that we do not know how to assign beliefs one by one, but rather that the very identity of a belief depends on its coherence with other beliefs, preferences, and so on. It makes no sense to be an atomist about propositional attitudes.

Davidson admits that physical theory and the network of propositional attitudes determine descriptive systems capable of representing a common range of events. Despite their common range, however, he argues that the disparate commitments of the physical and the mental reveal the systems to be genuinely distinct:

> It is a feature of physical reality that physical change can be ex-plained by laws that connect it with other changes and conditions physically described. It is a feature of the mental that the attribu-tion of mental phenomena must be responsible to the background of reasons, beliefs, and intentions of the individual. There cannot be tight connections between the realms if each is to retain its allegiance to its proper source of evidence. (97–98)

Evidence concerning physical theory must be described in physi-cal terms. Only then can the relation of an event to physical laws and to other events in its physical environment be shown; only under such a description can it enter into a physical explanation of the course of events. Conceived or described in any other way, that event would be evidence neither for nor against the theory in question. Correspondingly, evidence concerning the mental realm must be described in terms of propositional attitudes. Only then can its relation to background beliefs, intentions, and reasons be exhibited; only then can it enter into psychological or intentional explanations of the course of events. Even though the vocabularies proper to the mental and the physical refer to the same events, the disparate evidential and explanatory commitments of the two systems require that they provide distinct conceptions of those events.

By itself the argument that terms referring to mental events have systematically different criteria of application from terms referring to physical events does not entail the absence of a lawful connection between the two realms. Davidson contends, however, that it does so when conjoined with the principle of charity: We must impute coherence to the beliefs, desires, intentions, and actions we ascribe to an agent. "[W]hen we use the concepts of belief, desire, and the rest, we must stand prepared, as the evidence accumulates, to adjust our theory in the light of considerations of overall cogency: the constitutive ideal of rationality partly controls each phase in the evolution of what must be an evolving theory" (98). The attribution of rationality, and hence of a coherent system of beliefs, desires, preferences, and the like, is required to treat persons as persons. Davidson's position then is that explanations in terms of propositional attitudes are subject to a methodological constraint that is lacking in explanations in terms of physical characteristics. It is this methodological constraint, which stems from the concept of a person, that precludes a lawful relation between the mental and the physical.

The proper interpretation of the principle of charity is crucial. If no evidence could override the imputation of rationality, then the anomalism of the mental would surely follow. For whenever incoherence threatened we would be required to make ad hoc readjustments in our attribution of propositional attitudes. Because the content of each propositional attitude derives from its place in the system, the required readjustments would constitute subtle modifications in the criteria for the application of mental terms. Rather than recognize irrationality we would be required to revise not only our interpretation of the agent's beliefs, desires, and meanings, but also our understanding of what belief, desire, and meaning are. The dynamic character of mental types would obviously preclude their identification with static physical types.

But of course under certain circumstances we do call beliefs, preferences, and desires irrational. This makes the interpretation and status of the principle of charity problematic. To admit that it is sometimes reasonable to attribute an incoherent system of propositional attitudes to an individual is not necessarily to introduce global confusion into the realm of the mental. So long as we have a "background of true [or warranted] belief against which . . . failure can be construed" (96–97), it makes sense to recognize some degree of irrationality and error. One might argue, therefore, that on a more charitable reading, Davidson's principle would not require that the criteria for the application of mental terms be fluid enough to be modified each time any individual's system threatened to become incoherent. Rather, it would require that our concepts of belief, preference, intention, and the like be such that systems of propositional attitudes generally turn out to be reasonably coherent and rational and that, *ceteris paribus*, those concepts are to be applied in such a way as to minimize the incoherence and irrationality of each individual's set of propositional attitudes.

On this interpretation, the principle of charity seems to be simply a consequence of the holism of the mental. Since the content of each propositional attitude depends on its place within a system, without a reasonably coherent system we would be unable to identify particular propositional attitudes. And since a major motive for introducing propositional attitudes in the first place is to give explanations for actions in terms of reasons, without the imputation of rationality, we would fail at our assigned task. Thus to the extent that systems of propositional attitudes deviate from coherence and rationality, our faith in the particular identifications should be diminished.

Some such principle, however, is necessary to any holistic account. The introduction of a hypothetical entity or system of hypothetical entities is unjustified unless that entity or system actually serves the purpose for which it is introduced—unless it explains what it was introduced to explain. And whenever entities are identified on the basis of their place in a system, our knowledge of their existence and character is surely conditional on the coherence and explanatory power of the system. Accordingly, the claim that the principle of charity is a methodological rule peculiar to the realm of the mental appears unfounded. It is but an instance of the general (and uncontroversial) methodological requirement that hypothetical entities serve the explanatory purposes for which they are introduced.

Davidson's discussion of the holism of the mental does, however, argue forcefully that propositional attitudes must be conceived as hypothetical entities. Mental terms then purport to designate hypothetical entities. Their introduction as part of an explanatory scheme is vindicated, and the existence of their referents is (defeasibly) demonstrated if they lead to good explanations and predictions of behavior. The success of psychophysical generalizations made possible by their introduction is thus critically important. The systematic failure of definitional behaviorism demonstrates that there is no simple correlation between mental and behavioral predicates (92). Since an entire system of propositional attitudes must be postulated to account for intentional action—indeed, even to identify a given event as an intentional action—explanations employing mental terms cannot reasonably be construed as abbreviated descriptions of behavior. Rather, mental terms identify entities that are thought to be responsible for the production of certain behaviors, and it is because these behaviors can be accounted for by reference to the interrelation of beliefs, desires, preferences, and the like that they are identified as intentional actions.

According to Davidson, our current theory of the mental is heteronomic because events described in terms of propositional attitudes have causes or effects that are described without the use of mental terms. He concludes that to arrive at precise, exceptionless laws, our mental discourse would have to be replaced by (rather than just augmented by) the vocabulary of a comprehensive closed physical theory. The precise laws that underlie our psychological and psychophysical generalizations cannot be expressed in the same general vocabulary as those generalizations.

It is difficult to know how to evaluate this claim. Davidson does not explain how to determine whether proposed modifications belong to the same general vocabulary as the original generalization. Since his aim is to distinguish between the vocabulary of propositional attitudes and that of physical theory, one might suppose he means to employ the distinction between intensional and extensional discourse. That distinction, however, will not serve his purpose. Since laws, on Davidson's account, sustain counterfactual and subjunctive claims, they (and the explanations in which they occur) are as intensional as propositional attitude ascriptions are. Davidson gives no reason to suppose that homonomic generalizations and generalizations employing the vocabulary of propositional attitudes differ in logical form. Therefore his argument does not demonstrate that differences between the vocabulary of a homonomic generalization and that of a heteronomic generalization can be explicated by reference to the logic of the explanations in which they occur.

In any case, Davidson does take the question of psychophysical reduction to be a question of the relation between two distinct vocabularies: the vocabulary of the mental and the vocabulary of the physical. Because the distinction between heteronomic and homonomic generalizations is a distinction in the form that theoretical development will take, the growth of science is of central concern. Theoretical development involves the identification of entities described in terms dictated by one theory with entities described in terms dictated by another. Since such identifications are effected by correspondence rules, the proper interpretation of correspondence rules is crucial.

In order for our psychophysical generalizations to lend support to singular causal claims and related explanations of particular events, as Davidson maintains they do, the underlying laws must explain the success of those generalizations. Traditionally, such generalizations would be treated as empirical laws whose success is explained by their derivation from the underlying theory. Clearly such a procedure is unavailable to Davidson. If a generalization were derivable from theoretical laws, it would be homonomic; it could not contain predicates alien to the laws from which it was derived. Further, any such derivation would establish a type-type identity; events of the kind mentioned in the psychophysical generalization would be shown to be identical

with events of the kind picked out by the physical predicates of
the theoretical laws.

Wilfrid Sellars offers an alternative account of the success of
nontheoretical (or relatively nontheoretical) generalizations.[4] He
argues that the underlying theoretical laws explain the facts that
there are, hence derivatively explain why those facts conform
to the nontheoretical generalizations to the extent that they do.
On the surface Sellars's account seems ideally suited to explain
the success of psychophysical generalizations. But it cannot do
so within the framework of Davidson's program. The facts
and events that a theory explains are facts and events as described
in the vocabulary of that theory. In order to explain the success of
nontheoretical generalizations, there must be correspondence
rules that bridge the gap between nontheoretical and theoretical
vocabularies. But such correspondence rules render the original
generalizations homonomic by bringing their predicates into
the vocabulary of the theory. Thus the establishment of appropriate correspondence rules would render psychophysical generalizations homonomic. Indeed, at least where there were precise,
true psychophysical generalizations, such correspondence rules
would establish the identity of types of mental events with types
of physical events. Evidently it is a requirement of Davidson's
program that correspondence rules are not part of the closed
comprehensive physical theory that explains the occurrence of
events having mental descriptions. Without correspondence
rules, however, the underlying physical theory only explains the
occurrence of causal sequences that in fact have psychophysical
descriptions; it cannot explain their having the psychophysical
descriptions they do. But then the underlying physical theory
cannot explain the success of our psychophysical generalizations;
it will not show that those generalizations genuinely support
singular causal claims and explanations of particular events.

These considerations are not restricted to the case of
psychophysical generalizations. If Davidson is correct in distin-

---

[4] Wilfrid Sellars, "The Language of Theories," in *Science, Perception, and Reality*
(London: Routledge and Kegan Paul, 1963), 121.

guishing between homonomic and heteronomic generalizations, then because correspondence rules would render any domain homonomic, they cannot belong to any comprehensive closed theory. Therefore, the explanatory success of any truly heteronomic generalization cannot be explained by the underlying homonomic theory. Our faith that most of the generalizations that make up science and practical wisdom lend support to singular claims and to the explanations of particular events cannot, on the view Davidson is committed to, be supported.

Davidson takes the system of correspondence rules linking two theories to constitute a translation manual—a set of rules for translating from the vocabulary of one theory into the vocabulary of the other. He contends that it follows from the indeterminacy of translation that the reduced theory might be mapped onto the reducing theory in a variety of ways, each compatible with the total evidence. He concludes that there is no fact of the matter regarding which mapping establishes *the* reduction, no fact of the matter regarding which set of correspondence rules is correct.

The indeterminacy of translation thus appears to entail the anomalism of the mental. Because the system of propositional attitudes is an open system—because not every event that is the cause or effect of an event describable in the vocabulary of propositional attitudes is itself describable in that vocabulary—it must be extended or reduced to yield a comprehensive closed theory. But any attempted extension or reduction runs afoul of the problem of the indeterminacy of translation. Reduction of the network of propositional attitudes to neurophysiological theory, for example, requires that the events described as believings, desirings, and so on, be identified with the entities recognized by neurophysiology. Since there are potentially a variety of acceptable systems of correspondence rules, each yielding a different set of identifications, the selection of any one such system is ontologically arbitrary. Although considerations such as simplicity, intuitive appeal, ease of calculation, and the like might influence the choice of a system, there are no factual or ontological considerations that determine a uniquely correct choice.

The same problem arises when we attempt to expand or refine a theory. In the one case, equivalence principles are required to establish that the entities described as *F*s in the restricted theory are among the things (or are the vary same things) we describe as *F*s in the extended theory. In the other, such principles are required to show that what had formerly been described as *F*s are taken by the refined theory to be *G*s. Theory expansion or refinement then, like theory reduction, is subject to the indeterminacy of translation.

The foregoing discussion demonstrates, however, that the problem of radical translation cannot justify treating some developing theories as homonomic and others as heteronomic. If a system of correspondence rules constitutes a translation manual, then the relation between a theory and its successors is always mediated by analytical hypotheses. Hence, if radical translation is to be our guide, developing theories are all heteronomic.

Indeed the problems raised by this account are not restricted to developing theories. Because different sciences and different levels of complexity within a science are related by correspondence rules, if Davidson's account is correct, those relations render even finished science heteronomic. Using Davidson's criterion we are forced to recognize the anomalism of the chemical because we cannot establish a unique reduction of chemistry to physics, and even the anomalism of the physical, because there is more than one mapping of physical theory onto itself.

The anomalism of the mental is then a consequence of neither the special character of the mental nor the relation between reducing and reduced theories in science. Rather, it rests on the fact that whenever we attempt to establish that two descriptions are descriptions of the same thing, our accounts are vitiated by the inscrutability of reference and the resulting indeterminacy of translation. There is no fact of the matter with respect to which our identifications are correct or incorrect.

Indeterminacy, then, is a universal problem, infecting not only translation between radically disparate languages, but also ordinary discourse within a single language, and even the monologues of each individual speaker. But does indeterminacy raise special problems for theoretical reduction beyond those infecting all of language? I think not.

Relations between reducing and reduced theories need not and ought not be construed as a translation manual or a system of analytical hypotheses. Rather, correspondence rules are substantive claims of an empirical theory that comprehends both the reducing and the reduced theory. Whether pain is identical to c-fibers firing or heat is identical to mean molecular kinetic energy is a question of fact. Whether "gavagai" is equivalent to "rabbit" or to "undetached rabbit parts" is not. Correspondence rules, like the one linking pain to the firing of c-fibers, make factual claims. They can be true or false.

Nevertheless, the fact that the objects of one theory are identical with objects (or combinations of objects) of another, and that the predicates of the first are coextensive with predicates of the second is not sufficient to demonstrate that the former theory is reducible to the latter. For reduction to be justified, the two theories must be lawfully connected. The question then is this: Given that we have true (or at any rate, warranted) correspondence rules linking the two theories, what more is required to recognize them as lawlike?

Recall that when one theory is successfully reduced to another, the conceptual apparatus of the reduced theory is shown to be superfluous. Accordingly, that conceptual apparatus can in principle be eliminated *with no loss to science*. At least two questions must be answered in order to determine whether one theory can be eliminated in favor of another with no loss to science: the first concerns the growth of science; the second, the standards of adequacy for scientific explanations.

The question whether two theories make reference to the same objects and characterize them by means of coextensive predicates has a determinate answer only if theories are construed as systems of sentences. Reduction then is always reduction of theories as regimented by a canonical notation. But both the reducing theory and the reduced theory belong to growing, developing sciences. And as they grow and develop, the structure of their theories—as exhibited in their canonical forms—changes. If their common subject matter contains recalcitrant evidence, the theories must be modified to accommodate it.

Quine and Duhem have demonstrated that there is no unique correction dictated by either the structure of the theory or the structure of the evidence. If the two theories retain conceptual or methodological autonomy, then within the context of their respective research programs, alternative types of modification might be appropriate. For a variety of reasons, stemming from each theory's conception of its own domain, its emphasis, its methodology, its characteristic problems, and its values, alternative revisions might be called for. The scientific unity achieved by reduction is but one value and must be weighed against others in decisions regarding the rational preferability of different modifications.

Accordingly, although it is based on the relation between the sets of sentences that constitute our current theories, reduction has a prospective and regulative aspect as well. Since the result of reduction is the eliminability of the conceptual apparatus of the reduced theory, the recognition of correspondence rules as lawlike involves a commitment to the direction of scientific progress: the commitment that henceforth both sciences are to agree about which corrections are to be made to bring their common theory into accord with the evidence. With the elimination of the conceptual resources of the reduced theory, its basis for autonomous development is lost to science. In deciding whether the conceptual resources of one theory can be eliminated in favor of those of another then, our attention cannot be restricted to current theories. We must consider the prospects for scientific development as well.

At least part of what is involved in claiming that a theory can be eliminated in favor of another with no loss to science is that the explanations of the former can be replaced by explanations of the latter without loss. This clearly requires that the standards for adequate explanations be shared by the two theories. Otherwise, even if we demonstrate that the objects of the two theories are identical and the predicates coextensive, it is not clear that the correlation between the two should be taken as lawlike.

If the standard Hempelian account yields *sufficient* conditions for adequate explanations, this raises no problems. But if explanation is interest relative, and if two theories that treat of the same objects can yield explanations that serve *different* interests, then the reduction of one to the other will involve a loss of explanatory power. Although each explanation of the reduced theory will correspond to an explanation of the reducing theory, the latter will not be adequate to the purposes that the original explanations were designed to serve.

For example, it is reasonable to suppose that biology will recognize as laws certain generalizations concerning the goal-directed behavior of animals. According to Davidson, however, to treat human beings as persons is to rationalize their behavior. This involves treating their actions as anomalous. Thus, if Davidson's claim is correct, generalizations that function as laws in biology cannot be recognized as laws by the social sciences. Their status as law is suspended when the focus of interest is restricted to humans.

It is not because we believe *Homo sapiens* to constitute a singularity in the natural order that we refuse to take such biological generalizations as the basis for our explanations of human action. Rather it is because we recognize that our explanations of human action are subject to a requirement—that of rationalizing behavior—to which our accounts of the behavior of other animals are not. This is not to deny that there are contexts of inquiry in which it is appropriate to treat human goal-directed behavior on a par with the goal-directed behavior of other animals. But where our purpose is to rationalize behavior—as it is in the social sciences and generally in explanations in which human behavior is conceived as action—the true, warranted, universal generalizations in question are not to be recognized as laws.

Davidson explains the supervenience of the mental on the physical by saying that "there cannot be two events alike in all physical respects but differing in some mental respect, or that an object cannot alter in some mental respect without altering in some physical respect" (88). If my account is correct, then the supervenience of the mental on the physical amounts to this: Although every mental event is physical, it is not the case that every adequate explanation of it as physical is an adequate explanation of it as mental. And in general, science *A* supervenes on science *B* if the objects recognized by *A* are, or are wholly constituted out of, objects in the domain of *B*, but the standards for adequate explanations dictated by *A* are not shared by *B*.

Pessimism about the prospects of psychophysical reduction might be grounded in our recurrent failure to discover true or warranted psychophysical generalizations. Such pessimism is not unreasonable. But it does not represent Davidson's position. "The thesis is . . . that the mental is nomologically irreducible: there may be *true* general statements relating the mental and the physical, statements that have the logical form of a law; but they are not lawlike . . . if by absurdly remote chance we were to stumble on a nonstochastic true psychophysical generalization, we would have no reason to believe it more than roughly true" (90). Note what is being claimed here: although psychophysical generalizations may be true, none are lawlike. They must therefore be regarded as accidental truths—truths which establish de facto correlations between the mental and the physical, but which are logically too weak to support reduction. Further, were we to consider a true psychophysical generalization, we would never really believe it. Regardless of the weight of evidence supporting it, we would take it to be only approximately true.

Davidson claims that to treat an individual as a person, we must treat him as rational, and that to treat him as rational, we must describe his behavior in the vocabulary of propositional attitudes. He concludes that to admit the possibility of psychophysical laws amounts to changing the subject—"deciding not to accept the criterion of the mental in terms of the vocabulary of propositional attitudes" (90). That is, deciding not to interpret that behavior as the action of a person. Because the reduction of one theory to another justifies the elimination of the vocabulary of the former from the language of science, the reduction of the mental to the physical would justify the elimination of the vocabulary of propositional attitudes. By ceasing to identify mental events in the way that we used to, we change the subject.

The replacement of a given set of criteria by a more accurate or more useful set is, of course, a staple of scientific development. If this is changing the subject, then science changes its subject regularly. If the psychophysical case is an exception to this general practice, it is because the importance of the concept of a person makes the subject one we are particularly reluctant to change.

Davidson is, in effect, stipulating what it is to be a person. Stipulating criteria for membership in a kind is a common and (in general) sound semantic procedure. There is an aspect of this particular stipulation, however, that should be noted. Normally when a stipulation establishes criteria for inclusion in a kind, it is a question of fact whether anything satisfies those criteria. Since it is a question of fact, the methods of science are to be employed to discover the answer. Davidson contends, however, that science can tell us nothing about what individuals (if any) satisfy the criteria for being a person. Science is thus incapable of sustaining or overruling the employment of those criteria. Descriptions of behavior in terms of propositional attitudes are indefeasible by any possible scientific findings.

Davidson's position is this: even if we had a comprehensive closed physical theory and true psychophysical generalizations linking the system of propositional attitudes to that theory, to treat those generalizations as lawlike would be to give up treating individuals as persons. Can such a position be justified? Davidson's purely semantic account cannot do so. It can, to be sure, stipulate criteria for the application of the term "person" which make his position coherent. Since those criteria presuppose the anomalism of the mental, they guarantee that recognizing psychological laws amounts to changing the subject. But such an account cannot explain why we should go to such extremes to avoid changing the subject in question.

Epistemologically the situation is this: since Hume, it has been recognized that we are never epistemically obliged to draw a general conclusion from a finite base of evidence or to treat a generalization as lawlike. Choices between options are the result of two factors: the relative values of the two outcomes and the probability of those outcomes. Since a finite body of evidence never gives a general conclusion a probability of 1, the decision to accept a general claim on the basis of instances, and the decision to treat it as lawlike are made under uncertainty and risk. In deciding whether to recognize psychophysical generalizations as lawlike, we must weigh the benefits of succeeding in reducing the mental to the physical against the costs of falsely believing we have done so. If we refuse to treat psychophysical generalizations (however well confirmed) as lawlike, it is because we recognize that no matter how well confirmed our theory is, it might be false, and we estimate that the cost of mistakenly adopting a type-type identity theory would far outweigh the benefits to be gained from successful psychophysical reduction.

The weighting is, and I think must be, ideological. Central to the ideology underlying our moral and political theory is the view of a person as a rational, responsible agent. The notions of rationality and responsibility (obscure though they are) involve a recognition of the right of the individual to conceive of his behavior (and to have others conceive of it) as consisting of actions undertaken on the basis of his own beliefs, desires, and preferences.

To accept a type-type identity theory is to accept the view that each individual's behavior is determined by distributions of matter and psychophysical laws. This may be true. But if it is false, and we do not know how to reconcile free will with determinism, then in accepting it we commit the grave injustice of treating human beings as deterministic automata rather than as responsible agents. And it is the recognition of the magnitude of that potential injustice that makes us reluctant to accept psychophysical reduction. Given this evaluation of the seriousness of the risk, we would be not only morally, but also epistemically irresponsible to take it.

Our epistemic goals are achieving understanding and avoiding error. Where the two diverge, we must choose between them. If we deny that warranted psychophysical generalizations are lawlike, our primary concern is to avoid a particular error—the error of falsely believing the behavior of human beings to be fully determined by psychophysical laws and therefore wrongly denying that we are rational, responsible persons. In our choice whether to treat psychophysical generalizations as lawlike, the value of a unified scientific theory conflicts with ideological values. In this case, ideological values prevail.

The features of psychophysical reduction that I have emphasized are, I suspect, characteristic of reduction in general. The reduction of one theory to another is straightforward only if, in addition to satisfying the evidential and semantic requirements, the reducing and reduced theories share cognitive values. Where one theory merely supervenes on another, we cannot maintain that the one is reducible to the other without loss of explanatory power. For there is no guarantee that explanations acceptable from the point of view of the reducing theory are also acceptable from the point of view of the reduced theory, or that the interests served by the former will be served by the latter.[5] It will, to be sure, sometimes happen that the loss in explanatory power is outweighed by the benefits of a unified theory. In these instances reduction should be carried out. But it requires a normative argument to demonstrate that this is (or is not) the case.

In arguing that there is an ineliminably normative aspect to theory reduction, I do not mean to suggest that this aspect is subjective, or idiosyncratic, or arbitrary. It was precisely to avoid having to decide arbitrarily whether or not correspondence rules are lawlike that I was driven to appreciate the role of norms. The decision to recognize one theory as reduced to another (and therefore eliminable in favor of the other) must be grounded in good reasons if it is to be scientifically valid. If my account of reduction is correct, it is incumbent on the philosophy of science to discover what constitute good reasons for saying that a generalization ought to be regarded as lawlike. The arguments of Quine and Goodman show that the facts of the matter do not suffice to decide the case. I suggest that to solve the problem we must also look to the interests that the science is designed to serve, the sorts of explanations it intends to provide, the range of questions it seeks to answer, and the cognitive values it is committed to uphold.

[5] Philip Kitcher makes a similar point about biological reduction. He argues that the reduction of cytology to molecular biology fails because the processes that must be recognized as natural kinds in cytology cannot be recognized as natural kinds from the perspective of molecular biology. The theories then are heteronomic with respect to each other. See Philip Kitcher, "1953 and All That: A Tale of Two Sciences," in *The Philosophy of Science*, ed. Richard Boyd, Philip Gasper, and J. D. Trout (Cambridge: MIT Press, 1991), 558–559.

CHAPTER 3

Relocating Aesthetics:
Goodman's Epistemic Turn

Long a denizen of the realm of value theory, aesthetics emigrates, at Nelson Goodman's invitation, to epistemology. The arts function cognitively, Goodman insists. The job of aesthetics is to explain how. Such a contention would be capricious if epistemology were construed as the theory of knowledge. Rarely are the arts repositories of justified true beliefs. But knowledge, Goodman and I contend, is unworthy to be our ultimate cognitive objective. Far better to set our sights on understanding. And far better for epistemology to treat understanding as the focus of its concern.[1] In making a place for aesthetics in epistemology, Goodman thus reconceives epistemology as well as aesthetics. In so doing, he revitalizes both.

To understand a portrait, a partita, or a pas de deux, Goodman believes, is not to consider it beautiful, appreciate it, ascertain what its author intended by it, or have a so-called "aesthetic experience" of it. Rather, to understand it is to interpret it correctly—to recognize what it symbolizes and how it fits with or reacts against other world versions and visions.[2] Understanding

---

[1] Nelson Goodman and Catherine Z. Elgin, *Reconceptions* (Indianapolis: Hackett, 1988), 163.
[2] Nelson Goodman, *Ways of Worldmaking* (Indianapolis: Hackett, 1978), 109–140.

works of art is not a matter of passive absorption, but of active intellectual engagement with symbols whose syntactic and semantic features are often elusive. No more than in science is correct interpretation in the arts assured. Thinking you understand a symbol does not make it so.

Understanding a symbol may be difficult, for there are multiple modes of reference. Two are basic: denotation and exemplification. A symbol denotes what it applies to: a name denotes its bearer; a predicate, the objects in its extension. Goodman extends this familiar construal to accommodate nonverbal symbols. A portrait denotes its subject; a general picture, the members of the class it characterizes. Thus Manet's portrait of Berthe Morisot denotes Morisot, and a picture of a mallard in a bird watcher's guide denotes the members of the class of mallards. Depiction, according to Goodman, is pictorial denotation.[3]

Fictive symbols are denoting symbols, but they lack denotations. They derive their significance, Goodman urges, from certain terms that denote them. "Maggie-description" denotes the names and descriptions in *The Mill on the Floss* that conspire to fix Maggie Tulliver's fictive identity. "Unicorn-picture" denotes the portions of paintings, drawings, and tapestries that determine the constitution of the fictive kind, *unicorn*. "Ideal-gas-description" denotes the words and equations that determine the character of the fictive ideal gas.[4]

Abstract art does not even pretend to denote. Nor typically does architecture or instrumental music. Dance, too, frequently eschews denotation. Such works refer in other ways—often by means of exemplification. I. M. Pei's addition to the Louvre is a case in point. Because it both is a pyramid and presents itself as such, it exemplifies its shape.[5] Any symbol that at once instantiates and refers to a feature exemplifies that feature.[6]

Exemplification, like denotation, is ubiquitous. It links a sample

---

[3] Nelson Goodman, *Languages of Art* (Indianapolis: Hackett, 1976), 3–6.

[4] Ibid., 21–26.

[5] Nothing of philosophical substance turns on the correctness of my interpretation of any particular work of art. The reader who disagrees with my interpretations can easily supply examples of her own.

[6] Goodman, *Languages of Art*, 52–67.

to what it samples and an example, to what it is an example of. Exemplification is thus a staple of commerce, science, and pedagogy, as well as art. A free sample of laundry detergent exemplifies the soap's cleaning power; a blood sample, the presence of antibodies. A theorem exemplifies its logical form, while a sample problem in a textbook exemplifies the reasoning it seeks to inculcate.

A symbol can denote anything, so long as appropriate conventions are in force. Thus Pei's pyramid can denote my cat, if we establish a convention to that effect. But a symbol can exemplify only features it has. Not being a circle, Pei's pyramid is incapable of exemplifying circularity. Not being a cat, it cannot exemplify felinity. Exemplification, moreover, is selective. A symbol denotes everything it applies to, but exemplifies only some of the features it has. Even if Pei's pyramid was commissioned on a Tuesday, it does not exemplify *commissioned on a Tuesday*. For it does not highlight, exhibit, display, or convey that property.

Everyone uses examples, but few philosophers have appreciated their function. Examples are not merely decorative or heuristic devices, though they are often treated as such. They advance understanding in ways descriptions cannot. They show forth aspects of themselves, making those aspects available for exploration, elaboration, and projection. Wittgenstein and Kuhn extol examples and ground their philosophies in them. Goodman does more. He explains how examples function.

By exemplifying a feature, an example or other symbol affords epistemic access to it. Exemplified features need not be obvious. Often they are remarkably obscure. An intricate experiment may be mounted to exemplify minute differences in electromagnetic radiation. Mondrian's *Trafalgar Square* exemplifies astoundingly precise geometrical relations. The insight a work of art or a scientific experiment yields is seldom limited to a single case. Typically, it reverberates, as exemplified features and their kin turn up in other settings. A telling example opens a window on a world. That Goodman's discussions of exemplification occur almost exclusively in his works on aesthetics is perhaps unfortunate. The arts have no monopoly on the device. Although philosophy of science has yet to acknowledge it, without exemplification empirical science would be mute.[7]

Denotation and exemplification are not mutually exclusive. Works of art that denote usually exemplify as well. *Arrangement in Black and Gray* exemplifies shades of gray while denoting Whistler's mother. *War and Peace* denotes the Battle of Borodino while exemplifying Tolstoy's philosophy of history. That a single symbol can perform multiple referential functions is a central tenet of Goodman's aesthetics. Indeed, he takes multiple reference to be symptomatic of aesthetic functioning.[8]

Reference need not be literal, Goodman maintains. Symbols genuinely refer to the objects they figuratively characterize.[9] An indiscriminately enthusiastic undergraduate is genuinely, because metaphorically, a panting puppy. Brancusi's literally solid *Bird in Space* genuinely, because metaphorically, exemplifies fluidity; it both refers to and metaphorically instantiates the feature. The grue paradox genuinely, though not literally, pulled the rug out from under advocates of a syntactic solution to the problem of induction.

Truth is no more confined to the domain of the literal than reference is. If the student is metaphorically a panting puppy, "The student is a panting puppy" is metaphorically true. To be metaphorically true is to be true when interpreted metaphorically, just as to be literally true is to be true when interpreted literally. Nontautologous sentences are true only under an interpretation. Goodman's point is that when it comes to assigning truth values, whether the interpretation is literal or metaphorical makes no difference. Figurative reference then is no watered-down substitute. It performs all the symbolic functions of literal reference, and others besides.

Goodman's discussion of metaphor abounds with metaphors that exemplify the features he describes. In typically Goodmanian fashion he eschews literal characterization and describes metaphor's operation metaphorically. Thus, he contends, a metaphor

[7] See Catherine Z. Elgin, "Understanding: Art and Science," *Midwest Studies in Philosophy* 16 (1991): 196–208.

[8] Goodman, *Ways of Worldmaking*, 68.

[9] Goodman, *Languages of Art*, 68–85.

is "an affair between a predicate with a past and an object that yields while protesting".[10] His contention does double duty, both describing and exhibiting the interplay of attraction and resistance metaphor requires. Without resistance, a new application is literal; without attraction, it is arbitrary. Where an object both attracts and resists the application of a term, that application is metaphorical. Goodman's characterization needs no literal gloss. As my discussion amply illustrates, the temptation is not to paraphrase, but to elaborate—to see how much insight the description of metaphor as seduction will yield. By practicing what he preaches, Goodman both argues for and illustrates the tenability of his account.

Symbols do not ordinarily operate in isolation. They belong to, and function as members of families of alternatives that collectively sort the objects in a realm. "Panting puppy" belongs to a scheme that literally sorts dogs. Metaphor, Goodman maintains, exports the scheme to a distant realm, or repositions it to effect a novel sorting of its native realm. Thus the scheme that sorts dogs transfers to and effects a reorganization of people. Under that transfer, an enthusiastic undergraduate qualifies as a panting puppy; an unusually vicious critic, as a rabid Rottweiler; a trendy, self-promoting aesthete, as a prancing poodle. Novel patterns and distinctions reveal themselves as the metaphorical scheme sorts people into categories no literal scheme recognizes. Much of this is tacit. By calling one person a puppy, we make other dog labels available for characterizing people, whether or not we actually employ those labels.

Goodman's endorsement of metaphorical reference and truth connects with his nominalism. Contemporary realists are prone to think that literal language at its best partitions its domain into natural kinds, or divides nature at the joints, or discloses the true and ultimate structure of reality. Somehow, the world is supposed to dictate its proper description. Goodman denies this. He believes that any order we find is an order we impose. Systems of categories are contrived to impose order. They divide a domain into individuals and group those individuals into kinds. They

[10] Ibid., 69.

thereby equip us to describe, predict, explain, and complain about the entities thus recognized. But the success of one category scheme does not preclude the success of others. There is no unique way the world is, hence no privileged way the world is to be described.[11] A single domain may be organized in multiple ways; and for different purposes, different classifications may be best. Political geography and physical geography, for example, characterize their common domain quite differently, the one delineating the boundaries of cities and states, the other, the boundaries of forests and swamps. Each yields truths about the entities its terms refer to. Neither invalidates the other.[12]

Similarly, Goodman maintains, a literal and a metaphorical scheme may organize a common domain and yield divergent truths about it. No more than the adequacy of the terminology of political geography discredits that of physical geography does the adequacy of a literal scheme discredit that of a metaphorical one. To call a freshman enthusiastic is not to deny that he is a panting puppy.

Metaphor's cognitive utility is plain. A metaphorical application reorganizes a domain, sorting its constituents into hitherto unrecognized kinds, revealing novel kinships and differences. We could, of course, achieve the same reorganization by coining new literal terms. But first we would need to decide where the lines should be drawn. Metaphor saves us the trouble. It redeploys a partition that has already proven its worth. And its new deployment recalls and depends on its previous successes. For a metaphorical application, even if unprecedented, is not arbitrary.

There are, of course, no guarantees. But its prior effectiveness affords some reason to think a scheme will provide an illuminating classification of the constituents of the new domain. That the distinction between puppies and mature dogs is worth drawing in the canine realm suggests that it might mark a useful divide in other realms as well. When the application of "panting puppy" likens certain students to young dogs, when we come to see both groups as endearing and frustrating in much the same way, the

[11] Nelson Goodman, *Problems and Projects* (Indianapolis: Hackett, 1972), 24–32.
[12] Goodman, *Ways of Worldmaking*, 91–107.

suggestion is reinforced. Because "panting puppy" characterizes a class of students that no literal predicate captures, it enables us to see that those students have something in common that other students—even other enthusiastic students—do not share. Because it applies metaphorically, it likens the students it denotes to the literal referents of the term. Metaphor is a device for breaking through conceptual barriers. It affords epistemic access to novel affinities both within and between domains.

That metaphors can be true and illuminating does not, of course, mean that every metaphor is either. Some are simply false. That a lumbering lineman is a gazelle is no more true metaphorically than it is literally. Others, though true, are hackneyed. A knockout blow no longer packs much punch. Metaphor then is no sure-fire source of insight. Neither is literal language. But like literal language, metaphor affords an avenue to understanding.

Being inanimate, works of art cannot literally instantiate emotions. Since exemplification requires instantiation, they cannot literally exemplify emotions either. But they can and often do both instantiate and exemplify emotions metaphorically; and they can and often do instantiate and exemplify other metaphorical features as well. A literally lifeless sculpture may metaphorically exemplify liveliness and joy. A literally colorless, carefully crafted mazurka may metaphorically exemplify spontaneity and color. No more than denotation is exemplification restricted to the literal.

Expression, Goodman contends, is a type of metaphorical exemplification.[13] A work of art expresses aesthetic features it exemplifies. Thus, Michelangelo's *Moses* expresses barely controlled rage. The combination of opulence and decay in seventeenth-century Dutch still lifes expresses ambivalence about worldly success. Goodman denies that expression is restricted to the realm of emotion. Works of art metaphorically exemplify other aesthetic features as well. Bach's *Art of the Fugue*, for example, expresses symmetry and shape. There is evidently no a priori limit on the features art can express.

[13] Goodman, *Languages of Art*, 85–95.

Nevertheless, art does not express every feature it metaphorically exemplifies. A blocked writer's work in progress may be a perfect example of a metaphorical black hole, in that it absorbs her completely but returns nothing. Still, it would not express *black hole*, for it exemplifies the metaphor qua frustrated effort, not qua work of art.

The difficulty is to say what it is to exemplify *qua work of art*. Here Goodman is not as helpful as one might wish. He offers neither a criterion for art in general nor a criterion appropriate to each separate art. He says that a picture expresses only such properties as are constant relative to its literal pictorial properties. That is, it expresses only metaphorical properties that do not vary so long as the literal pictorial properties remain fixed.[14] But he says distressingly little about which of a picture's myriad properties qualify as literal pictorial properties.[15]

This, I suggest, is no accident. Goodman can give no exhaustive specification of literal pictorial properties, because we're still learning what they are. Moreover, we're learning, not primarily from aesthetics or art criticism, but from art itself. As new works exemplify new ranges of literal properties, we are made mindful of them. We realize that they were present and significant in earlier works as well. Thus we come to appreciate, as our predecessors did not, that the viscosity of paint, the weave of the canvas, the topography of the painted surface are literal pictorial properties. As a result, we acquire the capacity to recognize that metaphorical properties that are constant relative to these literal properties are expressed. This is a thoroughly Goodmanian conclusion, even though he never explicitly draws it.

Much reference is neither pure denotation nor pure exemplification, but a combination of the two. In allusion, Goodman maintains, a chain of reference consisting of denotational and exemplificational links connects a symbol with its referent. By portraying its subjects in costumes from a variety of historical periods, Rembrandt's *Night Watch* alludes to the proud history of the militia it depicts.[16] It denotes the figures garbed so as to

---

[14] Ibid., 86.

[15] Ibid., 42.

[16] E. Haverkamp-Begemann, *Rembrandt: The Night Watch* (Princeton: Princeton University Press, 1982), 84–93.

exemplify the militia's history, and thereby refers indirectly to that history.

Artists don't work in a vacuum. Their works often betray a host of influences. But influence is not the same as allusion. Raphael's *School of Athens*, for example, shows the influence of Michelangelo's Sistine Chapel frescoes, but does not allude to them. Carravagio's *Calling of Saint Matthew*, on the other hand, makes the allusion. For Christ's gesture to Matthew in the Carravagio harks back to and derives its authority from Michelangelo's portrayal of God reaching out to Adam.[17] The difference is this: Although the Raphael exemplifies features it shares with and takes from Michelangelo's frescoes, it does not use the shared features as a vehicle for referring to the frescoes. The Carravagio uses the features it shares as a bridge to (and from) Michelangelo's work. Allusion and other modes of referential action at a distance require not just the existence of intervening referential links, but that reference be transmitted across those links.

Variation, Goodman argues, is a complex form of indirect reference.[18] A variation must resemble its theme in some respects and differ from its theme in others. But every two passages do that, and not every passage is a variation on every other. What makes for variation, Goodman contends, is not just commonality and contrast, or even a specific sort of commonality and contrast, but reference via commonality and contrast. A variation refers to its theme via literal exemplification of shared features and via metaphorical exemplification of contrasting features. Only if reference is transmitted along both chains does a symbol's relation to another qualify as a variation on it. If Goodman's explication is correct, variation is not confined to music, for a symbol's status as a variation turns on its referential function, not on its aesthetic medium. The requisite functions can be performed in any art. Picasso's take-offs on *Las Meninas* and on *Le Déjeuner sur l'Herbe* then qualify as genuine variations.

Goodman's account provokes and provides resources for rethinking works of art we do not typically consider variations, for

[17] H. W. Janson, *History of Art*, 2d ed. (Englewood Cliffs: Prentice Hall, 1982), 484.
[18] Goodman and Elgin, *Reconceptions*, 66–82.

seeking out hitherto neglected relations within and between them. The variation form may be more prevalent than we know. The account also suggests further avenues for exploration. It invites us to investigate whether other aesthetic categories admit of explication in terms of indirect reference. It would be surprising if variation were in this respect unique.

Scientific symbols typically symbolize along comparatively few dimensions; aesthetic symbols, along comparatively many. The same configuration could serve as a symbol of either kind. A wavy line might function as an electrocardiogram, Goodman suggests, or as a Hokusai drawing.[19] Only its shape matters when it functions as an EKG. But when it functions as a drawing, the precise color and breadth of the line at each point, each particular shade in the background, the exact position and dimensions of the line on the paper, the paper's weight, composition, and texture—all are potentially significant.

Like other scientific symbols, the electrocardiogram is referentially austere. It denotes a series of heartbeats and exemplifies a range of symptoms. That's all. The drawing, on the other hand, performs multiple complex and interanimating referential functions. Through denotation, exemplification, expression, and allusion, it affords epistemic access to diverse referents by a variety of routes. Scientific symbols are comparatively attenuated, Goodman maintains. Aesthetic symbols are relatively replete.[20]

Moreover, the dimensions along which a scientific symbol symbolizes are ordinarily settled in advance. A cardiologist could discover that small irregularities in the cardiogram's curve are indicative of a subtle coronary malfunction. But she's unlikely to find that the line's intensity has any cardiological significance. That the once black line fades off to a pale gray indicates that the printer needs more toner, not that the heartbeat is weakening.

The drawing is more open ended. Despite its familiarity, we might easily discover that hitherto unacknowledged aspects function symbolically. A sensitive critic could come to realize that an

[19] Goodman, *Languages of Art*, 229.
[20] Ibid., 229–230.

almost indiscernible asymmetry in the paper's weave contributes to the picture's flow. This is another reason why works of art merit and reward repeated reading.

A portrait portrays Virginia Woolf, her head jauntily cocked. It can capture the exact tilt of her head, the exact line of her brow, for pictorial precision admits of no limit. Pictures are, in Goodman's terms, semantically dense. They belong to symbol systems that can reflect the finest differences in their referential fields. Does the picture simply depict Woolf? Or does it depict her looking relieved but slightly perplexed, or happy but mildly surprised, or bewildered but on the whole, content, or what? There may be no telling, for the referents of these and kindred characterizations may differ beyond the threshold of discrimination. To determine firmly and finally just what a given work depicts can be impossible. For pictures are semantically nondisjoint. It is not always possible to distinguish divergent referents.

Words too are semantically dense and nondisjoint. Language has the resources to describe an item with any degree of precision, and linguistic descriptions are so related that it is sometimes impossible to tell their referents apart. The difference between verbal and pictorial symbols, Goodman contends, lies in their syntax. Languages have alphabets—distinct and discriminable characters that compose their symbols. As a result, language admits of a criterion of syntactic equivalence—sameness of spelling. Inscriptions in a language that are spelled the same are interchangeable without syntactic effect.

Pictorial systems lack alphabets. They are syntactically dense. The exact color, thickness, position, and shading of each line in a drawing is critical to its identity as a pictorial symbol. Any two marks display some difference in these respects. So none are syntactically equivalent. No pictorial mark can replace any other without altering the symbol's identity.[21]

Computer graphics, television images, mosaics, and the like might seem to present counterexamples to Goodman's claim. Television images and computer graphics are generated by arrays

[21] Ibid., 130–154.

of digitally encoded dots. The dots are close together. But it is not the case that between any two there is a third. A mosaic consists of discrete tiles in a limited number of colors, sizes, and shapes that, like the computer's dots, seem to serve as an "alphabet"—a system of repeatable basic units that make up the picture. That computer pictures, television images, and mosaics are genuine pictures is beyond dispute. That they consist of discrete, discriminable syntactic elements is not.

Syntax is determined not by physical constitution, but by what constitutes an item as a symbol. To construe an item as a particular symbol is to classify it against a background of alternatives. The symbol together with its alternatives constitutes a symbol scheme. Each element obtains its syntactic character from its place in the scheme. The same mark may belong to several schemes, hence constitute several symbols. A mosaic pattern or a dot matrix design easily fits into a digital scheme—one whose characters are discrete and discriminable. But to construe them as pictures is to read them differently.

When we read a computer printout as a picture, we treat the array of grays that compose it as drawn from the full range of possibilities. Any shade of gray, it seems, could have been used. When we read a mosaic as a nativity scene, we treat its colors, sizes, and shapes as elements of a dense field of alternatives. Even if the artist was in fact limited in the choices available to him, we read the work as part of a scheme that provides unlimited options. Evidence for this can be found in our critical appraisals. When, for instance, we recognize the mosaics at Ravenna as masterpieces, when we say that the mosaicists got them exactly right, we mean that no conceivable alternative would have been better, not just that they are as good as can be expected given the limited options available. The schemes that constitute symbols as pictures thus provide for alternatives that discrete, discriminable characters cannot comprise. So when computer printouts, television images, or mosaic designs are construed as pictures, their material atoms—individual dots or tiles—do not function as their syntactic primitives. As much as paintings or drawings, such pictures are syntactically dense symbols.

Density and repleteness are not necessary for art; nor are articulation and attenuation required by science. But, Goodman contends, density and repleteness are symptomatic of art.[22] And, I would add, articulation and attenuation, symptomatic of science. The reason, I suggest, lies in the cognitive values science and art embrace.[23]

Science values reproducible results and intersubjective accord. It structures itself and its subject matter to achieve these ends. If its findings belonged to a dense, nondisjoint field of alternatives, there would be no way to tell precisely what they were, hence no way to tell whether they had been reproduced. If, for instance, every difference in temperature constituted a different finding, we could never claim that the temperatures of two samples were the same. For identical readings might mask divergences beyond the threshold of measurement. Science then has an incentive to partition its domain into discrete and disjoint alternatives. It restricts its parameters and counts its measurements accurate only to a specifiable number of significant figures. Science also has reason to reject repleteness. For if scientific symbols were replete, there would be no way to tell whether differences in some seemingly irrelevant respect were in fact significant. In reading a thermometer, for example, we could not safely ignore such features as the thickness of the column of mercury, the shape of the tube that contains it, or its distance from the magnetic north pole. We could then never settle what the instrument reveals. To be sure, science retains the option of increasing its precision, refining its categories, and increasing the range of factors it considers. But it cannot opt for absolute precision and unlimited range without abandoning hope of agreement among investigators and reproduction of results.

Art has different aims. It values sensitivity more highly than accord, and aspires to results that cannot be reproduced. It considers apparently interminable disagreements in interpretation a fair price to pay.

---

[22] Goodman, *Ways of Worldmaking*, 67–68.
[23] Catherine Z. Elgin, *With Reference to Reference* (Indianapolis: Hackett, 1983), 120.

> Where we can never determine precisely just which symbol of a
> system we have or whether we have the same one on a second
> occasion, where the referent is so elusive that properly fitting a
> symbol to it requires endless care, where more rather than fewer
> features of the symbol count, where the symbol is an instance of
> the properties it symbolizes and may perform many interrelated
> simple and complex referential functions, we cannot merely look
> through the symbol to what it refers to . . . but must attend con-
> stantly to the symbol itself.[24]

Ambiguity, vagueness, and equivocality are scientific vices; they
are often aesthetic virtues.

That the mind is mostly passive in the reception of sen-
sations Goodman emphatically denies. The mind, he insists is
always active—ceaselessly searching, discriminating, integrating,
and organizing. Nor, he insists, are our perceptual capacities
invariant. To the oft put allegation that for anyone who can
not distinguish an original from a forgery, the distinction
makes no difference, Goodman replies: What you can not
distinguish today, you may learn to distinguish tomorrow.
Further exposure to art can remedy even longstanding aesthetic
incapacities.[25]

Refinement of the sensibilities is not just a matter of making
ever more delicate sensory discriminations. It also involves developing new recognitional capacities, and new ways of structuring
the perceptual field. Learning to see an equivalence between a
vivid red and a vivid green is as much a perceptual advance as
learning to see the difference between scarlet and vermilion. Nor
does refinement of visual perception restrict itself to the realm of
colors. We learn to recognize patterns, styles, treatments, subjects,
and much more. Just by looking, a connoisseur can tell that a
painting is a Raphael, that it was painted before the artist encoun-
tered Michelangelo's work, that it achieves a harmonious balance
of color and shade, that it expresses serenity. The novice, survey-
ing the same work from the same vantage point, cannot yet see

[24] Goodman, *Ways of Worldmaking*, 69.
[25] Goodman, *Languages of Art*, 99–112.

these things. What we see depends on more than where we stand. Features that elude the casual viewer leap immediately to the expert's eye. Experience, habit, interests, and expectations inform perception. So over time, with effort and education, we develop an ability to see what we once could not.

Although Goodman recognizes no distinctively aesthetic emotion, he appreciates the role of emotions in the arts. Many works, we have seen, express emotions. We could hardly begin to understand such works were we oblivious to the emotions they express. This is fairly uncontroversial. More radical is Goodman's view that one's own emotional responses are vehicles for understanding. That a work amuses me is some reason to think it funny; that it bores me, some reason to think it banal. To be sure, my reaction does not entail that the work is funny or banal. Neither does its seeming blue entail that it is blue. But insofar as I am a good judge of color and conditions for judging color are propitious, something's seeming blue to me is a good, albeit defeasible, reason to believe that it is blue. Likewise, insofar as I have a good sense of humor and conditions are propitious, something's seeming funny to me is a good, albeit defeasible, reason to believe that it is funny.

Where the arts arouse emotions, conditions tend to be propitious. Overpowering emotions like abject terror, blind rage, or rapt infatuation do not typically present themselves as occasions for inquiry. Ordinarily they call for action, not contemplation. No one in the grip of genuine terror is likely to use her emotion as a scalpel for dissecting fear and its object. But emotions excited by the arts are muted and displaced. It is possible and may be informative to use the terror a Hitchcock film excites as a source of insight. We may find that modulations in our fear correlate with significant features of the film—features we would otherwise have overlooked. And we may learn to detect in ourselves subtle emotional nuances that we had previously lumped indiscriminately together under the label "fear". The insights we thus glean typically project beyond the aesthetic realm. We recognize the newfound nuances in our emotional responses and use those nuances as detectors of hitherto unrecognized aspects of the objects that occasion them. If Goodman is right, emotion is not the end of aesthetics, but a powerful means by which art advances understanding.[26]

Merit too converts from end to means. Rather than understanding art in order to evaluate it, we should, Goodman maintains, evaluate in order to understand. An unexpected assessment kindles curiosity, prompting us to attend more carefully to a work—to seek out and perhaps to find aesthetically significant features we had missed.[27]

But what makes for merit? Not beauty; many great works of art are ugly. Not truth; fictions are literally false, and works in the nonverbal arts neither true nor false. Rather, Goodman believes, aesthetic merit depends on rightness of symbolization. Rightness in turn depends on fit—"fit to what is referred to in one way or another, or to other renderings or to modes and manners of organization".[28] A work that easily fits is readily intelligible. We can tell what it refers to, how it characterizes its referents, and how it relates to other works. But fit alone is not enough. A mundane representation of a routine subject in a popular style fits all too well with other renderings and with familiar modes and manners of organization. That's what's wrong with it. It discloses nothing new. By its difficulty fitting in, a revolutionary work challenges familiar modes and manners of organization. It provokes a reconception of matters we thought were settled, perhaps by extending or reconfiguring the referential field, or by employing novel symbols or familiar symbols in novel ways. Such works are not so readily intelligible.

Revolutionary works, of course, are not entirely alien. They draw on as well as violate established conventions. For all its novelty, cubism owes a great deal to Cézanne. Revolutionary works too strive to fit, Goodman maintains. But lacking the luxury of nestling comfortably into a preestablished niche, they must adjust the background assumptions to create a space for themselves. The task is more difficult; the rewards make it worth our while. Such works advance understanding by disclosing fea-

---

[26] Ibid., 245–255.
[27] Goodman, *Problems and Projects*, 120–121.
[28] Goodman, *Ways of Worldmaking*, 138.

tures their predecessors masked, by revealing new ways to see and new things to find in our worlds.

Merit derives then not just from fitting, but from fitting and working: fitting with what we already understand and working to advance understanding. That is, "achieving a firmer and more comprehensive grasp, removing anomalies, making significant discriminations and connections, gaining new insights".[29] Of course, revolutionary works are not the only ones that advance understanding. Works that are firmly grounded in an entrenched tradition can wring new insights from it. To say that Mozart's late quartets exemplify established classical forms is hardly to fault them. Such works advance understanding by uncovering a tradition's previously untapped (and often unsuspected) powers, making novel and effective applications of its symbolic resources, deploying its devices with sensitivity and courage to illuminate what had been obscure.

Our cognitive objectives themselves evolve as new opportunities arise. Each new level of understanding provokes new questions, poses new problems, pushes inquiry in directions its predecessors could not have imagined. Picasso solved aesthetic problems Rubens lacked the resources even to pose. Understanding, as Goodman and I conceive it, is not a repository of fixed and final epistemic achievements, but a springboard for further inquiry. This is so in the arts as much as in the sciences.

A problem remains. Advancement of understanding is a standard of cognitive merit in general. If, as Goodman contends, the arts function cognitively, it applies to works of art. But are all works of art that advance understanding *aesthetically* valuable? Robert Nozick thinks not.[30] Had Newton expressed his laws in doggerel, the poem would have advanced understanding considerably. It would then have been cognitively valuable. But doggerel—even in the service of science—lacks aesthetic merit. Thus, Nozick concludes, aesthetic merit is not cognitive merit. For despite its contribution to physics, Newton's poem, as Nozick imagines it, would have been lousy art.

[29] Goodman and Elgin, *Reconceptions*, 158.
[30] Robert Nozick, "Goodman, Nelson on Merit, Aesthetic," *Journal of Philosophy* 69 (1972): 783–785.

What Nozick's criticism overlooks is that a symbol can function in several ways at once, and can function well in one way while functioning badly in another. Newton's doggerel would function well as science, badly as art. And only insofar as a symbol is functioning aesthetically, does its contribution to the advancement of understanding qualify as aesthetic merit.[31] The crucial question for Goodman then is not "What is art?" but "When is art?"[32]

Goodman ventures no real definition of art, no set of necessary and sufficient conditions on the aesthetic. This omission stems, I think, not so much from his qualms about analyticity as from his conception of aesthetics as a branch of epistemology. Goodman has always been less interested in closing the borders of the aesthetic to interlopers than in discovering epistemically significant affinities that cut across realms. Determining whether political cartoons, Navaho blankets, or handmade quilts qualify as art is not so important as understanding what and how they contribute to cognition, and how their functioning resembles and differs from that of related symbols in the arts and elsewhere. The demand for a demarcation criterion no longer seems pressing.

Goodman does not, of course, seek to reduce art to science or science to art. Rather he sees both as contributing to a general project of advancing and deepening understanding.

[31] Nelson Goodman, *Of Mind and Other Matters* (Cambridge: Harvard University Press, 1984), 138–139.
[32] Goodman, *Ways of Worldmaking*, 57–70.

CHAPTER 4

Facts That Don't Matter

Responsibility for the indeterminacy of translation is usually assigned to Quine's behaviorist assumptions. Quine prides himself on the connection.[1] Chomsky, Searle, and others castigate him for it, charging that indeterminacy amounts to a *reductio ad absurdum* of linguistic behaviorism.[2] But despite this broad consensus, it is unwise to assume that indeterminacy and behaviorism are so intimately related. Arguments developed by Hilary Putnam show that indeterminacy of translation and its consequences, inscrutability of reference and ontological relativity, survive the repudiation of behaviorist restrictions.

This resilience is unexpected, for univocality seems easy to achieve. The more exacting a standard, the fewer and more uniform its compliants. To foreclose the possibility of mutually incompatible translation manuals seems only to require a suitably demanding standard for translation. Easier said than done. It

[1] W. V. Quine, "Indeterminacy of Translation Again," *Journal of Philosophy* 84 (1987): 5.

[2] See Noam Chomsky, "Quine's Empirical Assumptions," in *Words and Objections*, ed. Donald Davidson and Jaakko Hintikka (Dordrecht: Reidel, 1969), 53–67; John Searle, "Indeterminacy, Empiricism, and the First Person," *Journal of Philosophy* 84 (1987): 124–127.

follows from Putnam's arguments that even criteria expressly designed to eliminate indeterminacy are not up to the task.

Putnam does not take his arguments to have this effect; he constructed and employed them to serve other ends. My goal here is to show that they support and extend the indeterminacy thesis. The endurance of indeterminacy in the face of sustained efforts to eradicate it suggests that it is integral to language. If so, it must be accommodated, rather than repudiated, by any adequate semantics.

Quine's detractors do not dispute the accuracy of the behaviorist's findings or the reliability of his methods. They contend, however, that the methods are unduly restrictive, capable of disclosing but a narrow range of relevant facts. There is, they insist, more to a language than meets the ear. The problem is to determine what more there is, and to secure epistemic access to it. That achieved, they maintain, translation will be determinate. This is doubtful. If Putnam is right, translation manuals that answer to all identifiably relevant facts can still conflict.

Quine, of course, purports to have shown this already. For he contends that all identifiably relevant facts are facts about verbal behavior and dispositions to verbal behavior. And his discussion plainly accommodates these. Still, his methodological parsimony may render his demonstration suspect; for scruples such as his might easily obscure facts that bear on the case. Putnam, though, is generous to a fault. He allows appeal to intentions and intensions; to functional states and possible worlds; to essences, properties, and causal powers. But even Putnam's prodigality is not unbounded. He insists that our theory, however extravagant its apparatus, must be consonant with actual language use. It may supplement or underwrite, but it cannot conflict with usage. This seemingly modest requirement turns out to be far more restrictive than anyone originally supposed.

The problem of radical translation arises when a linguist sets out to construct a translation manual for a totally alien tongue.[3] Hav-

---

[3] W. V. Quine, *Word and Object* (Cambridge: MIT Press, 1960), 26–79.

ing nothing else to go on, he must glean the meanings of native utterances from observed verbal behavior.

He starts by correlating utterances with conspicuous local events, translating them as sentences we would utter or accept in the circumstances. "Gavagai!", uttered as a rabbit hops by, is tentatively translated as "Rabbit!" Translations are offered provisionally. Further investigation may convince the linguist that "Jack rabbit!" or "Cottontail!" translates "Gavagai!" better than "Rabbit!" does. By trial and error, he zeroes in on a sentence we would accept when and only when similarly situated natives accept "Gavagai!" Such a sentence has the same stimulus meaning as "Gavagai!"

Normally a speaker will assent to any sentence he is prepared to volunteer. So, having discovered the stimulus meanings of some native sentences, the linguist can identify native signs of assent and dissent by seeing how his informants respond to native sentences he volunteers under plainly appropriate and under plainly inappropriate stimulus conditions. Then, by treating those signs as rough approximations of "true" and "false", he can identify truth-functional connectives, stimulus synonymy, stimulus analyticity, and stimulus contradiction.

But the marks of stimulus analyticity and stimulus contradiction, invariable assent and dissent, are unspecific as to subject matter. And stimulus synonymous sentences, though they coincide in circumstances of acceptability, need not agree about anything else. "George Bush is vice president" and "Ronald Reagan is president" are stimulus synonymous for anyone disposed to accept both if either, and to reject both if either. So stimulus synonymy does not establish sameness of meaning or reference. It does not afford a basis for correlating native terms with English ones, any more than stimulus analyticity or stimulus contradiction do.

Nor does stimulus meaning. The correlation of his locution with "Rabbit!" does not demonstrate that the native is committed to the existence of rabbits. For concurrence of verbal response does not demonstrate agreement about what is being responded to. So from their agreement, it does not follow that the linguist and his informant are, or take themselves to be, talking about the same thing.

Thus far, the linguist's problems have been purely inductive, a business of inferring generalizations on the basis of limited evidence. Further investigation may disconfirm his findings. Generalizations that are never disconfirmed may yet be false, since they can conflict with usages he happens never to encounter. Such is the scientist's lot. Still, as much as any other empirical scientist, the linguist is entitled to confidence that his highly confirmed results reveal facts of the matter—the actual verbal dispositions of speakers of the language.

But induction can take him no further. To identify native terms and their referents, to settle questions of individuation, the linguist must resolve native sentences into repeatable components and map those components onto his own words. This requires resort to analytical hypotheses.

To be adequate, a system of analytical hypotheses must yield translations that conform to linguistic behavior. The difficulty is that this requirement is too easily met. "Rival systems of analytical hypotheses can fit the totality of speech behavior to perfection, and can fit the totality of dispositions to speech behavior as well, and still specify mutually incompatible translations of countless sentences insusceptible of independent control."[4] That being so, Quine concludes, nothing favors one of the rivals over the rest. There is no fact of the matter.

Quine may seem a bit hasty here. The linguist's inability to determine more of the language without recourse to hypotheses does not show that there is nothing more to determine, or that his hypotheses are "analytical" rather than substantive. So we might, like Chomsky, take a translation manual to be an empirical theory that seeks to explain linguistic behavior by means of conjectures about underlying, psychologically real, semantic facts. In that case, a translation manual is correct just in case it answers to those facts.[5] Since theories in linguistics, as in any science, are

---

[4] Ibid., 72.
[5] Chomsky, "Quine's Empirical Assumptions," 53–67.

underdetermined by evidence, it is no surprise that conflicting, evidentially adequate translation manuals can be constructed. Nevertheless, Chomsky believes, the semantic facts, accessible or not, make one manual right, the others wrong.

Were the field linguist's predicament just an occupational hazard, Chomsky would be on strong ground. For we have found no reason to think theories of language less factual than other theories. But the problem of radical translation begins at home. In learning their native tongue, children have no more to go on than the linguist does. Like him, they resolve complex linguistic constructions into repeatable elements and recombine those elements to generate new constructions. Their efforts, like his, are tested against the verbal behavior of competent speakers. And they count as competent when their utterances conform to community standards.

If a linguist could go wrong through misidentifying or misclassifying the components of a language, so could a seemingly fluent native. Her error, like the linguist's, would be undetectable, since she shows every sign of linguistic competence. But her failure to resolve utterances of her interlocutors into, and construct her own utterances out of, the language's real elements would mean that she does not speak the language. The impeccability of her verbal behavior would not reduce the charge against her.

Surely this conception of competence is untenable. To speak English requires no more than consistently speaking as English speakers do. And however idiosyncratic her processing, our native, by hypothesis, does that. Nor can translation be expected to delve deeper than competence does. There is no fact of the matter of meaning beyond what linguistic behavior discloses, because meaningful speech is nothing more than good linguistic behavior. Language is deeply superficial.

But if there is no fact of the matter about what my language means, there is no fact of the matter about what I mean in speaking my language. For I can mean only what my language equips me to mean. It follows that whether I mean "rabbit" or "rabbit stage" when I say "rabbit" is indeterminate; whether I refer to rabbits or rabbit stages, inscrutable; whether I am committed to the existence of rabbits or rabbit stages, absolutely unintelligible.[6] A more counterintuitive result is hard to imagine.

What is counterintuitive is not necessarily wrong. To undermine Quine's conclusion requires discrediting his argument. Its defect, Searle contends, lies in its mistaken adherence to a requirement of publicity: by restricting the data base to publicly certifiable evidence, Quine deprives the linguist of a rich lode of semantic information—namely, what each of us knows from his own case.[7]

According to Searle, it is perfectly plain to me that by "rabbit" I mean "rabbit", not "rabbit stage" or "undetached rabbit part". And such plain facts demonstrate that indeterminacy does not infect my idiolect. Moreover, like cases should be treated alike. So, barring evidence to the contrary, I am entitled to infer that it is equally plain to every other speaker of my language that by "rabbit" he means "rabbit" too. Since a language consists of the idiolects of its speakers, the validity of such inferences secures determinacy of meaning and reference across the home language. In English then, "rabbit" means "rabbit" and refers to rabbits. Finally, it is plain to me that by "lapin" I mean "rabbit". Since like cases should be treated alike, I can infer that such first-person facts are equally plain to other translators. It follows that translation is determinate. Any bilingual is equipped to refute the indeterminacy thesis by generalizing from his own understanding of the languages he speaks, appealing only to the uncontroversial principle that like cases should be treated alike.

If Searle is right, indeterminacy of translation, inscrutability of reference, and ontological relativity derive entirely from an impoverished methodology that unjustifiably prohibits appeal to plain and plainly relevant psycholinguistic facts. So Searle maintains what Quine denies: that meanings are psychologically real and are epistemically accessible to the agent via introspection.

Such disagreements are notoriously difficult to resolve, for each party is convinced that his adversary begs the question or misses the point. Without a shared basis for settling disputes, positions

---

[6] W. V. Quine, "Ontological Relativity," in *Ontological Relativity and Other Essays* (New York: Columbia University Press, 1969), 47–48.
[7] Searle, "Indeterminacy, Empiricism, and the First Person," 123–146.

may harden, and argument disintegrate into name calling across an ever-widening abyss of mutual incomprehension. So rather than pit Quine directly against Searle, I want to consider whether Searle's method, if we use it, is sufficient to generate his conclusion. Arguments drawn from Putnam show that it is not.

Imagine a planet, Twin Earth, exactly like Earth but for the absence of H_2O and the presence in like quantities of XYZ, a complex chemical with water's phenomenal properties and ecological functions.[8] So similar are the two chemicals in their manifest characteristics that they can be distinguished only by sophisticated scientific tests. Suppose further that on Twin Earth, each of us has a nearly identical twin who speaks Twenglish, a dialect very close to standard English. In Twenglish, however, XYZ is called "water"; H_2O is not.

My twin and I are ignorant of chemistry, so our uses of "water" are indistinguishable. "Water" functions in her deliberations, conversations, and responses to the environment just as it does in mine. Moreover, owing to our ignorance, our psychological constitutions are indistinguishable as well. The plain facts that she introspects are no different from those I introspect. Each of us can sincerely affirm,

> By *"water"* I mean "water".

But she does not refer to H_2O; nor I to XYZ. Although our psychological states are indistinguishable, the referents of our terms are not. Reference then is not determined by what is in the head. So the plain facts Searle invokes are powerless to disclose it.

Nor can they fix meaning, as another divergence in dialects demonstrates. Twenglish speakers use "beech" when English speakers use "elm", and use "elm" when English speakers use "beech".[9] So Twenglish speakers would say

> The beech was struck by lightning

[8] Hilary Putnam, "The Meaning of 'Meaning,'" in *Mind, Language, and Reality* (Cambridge: Cambridge University Press, 1975), 223.
[9] Putnam, "The Meaning of 'Meaning,'" 226–227.

when English speakers would say

> The elm was struck by lightning.

I cannot tell a beech from an elm; nor can my twin. My twin and I know vaguely that both are large, deciduous trees and realize that they are trees of different kinds. But nothing in our understanding of the words or their objects differentiates beeches from elms. Nevertheless, my twin and I comprehend beech-talk and elm-talk, and use the terms competently—if uninformatively—in daily life.

Because of our ignorance, my "idea" of a beech is the same as my twin's; my disposition to use "beech", the same as hers. Each of us can sincerely affirm

> By "*beech*" I mean "*beech*".

Nevertheless, our meanings differ. For my affirmation is in English; hers, in Twenglish. Nor is the fact that we speak different languages disclosed by differences in our psychology. For English and Twenglish coincide in our idiolects. They diverge only in the usage of our more knowledgeable compatriots. The first-person perspective avails me nothing; the plain facts it purports to disclose neither determine what I mean nor afford me a secure basis for generalization. The moral of Putnam's fiction is this: we often do not know what we mean or what we are talking about, but our linguistic competence is none the worse for that.

Introspection thus cannot disclose mental determinants of meaning. But science sometimes reveals what self-scrutiny conceals. So before abandoning the search for meaning in mind, we should consider the claims of cognitive science.

Fodor contends that the meaning of a word derives from its mental counterpart, a symbol in the "language of thought".[10] Locutions answering to the same mental representation are alike in meaning and in reference, and perform the same functions in their respective languages. Locutions whose meanings and refer-

[10] Jerry Fodor, *The Language of Thought* (Cambridge: Harvard University Press, 1975), 79–97.

ents differ correspond to distinct mental representatio. is and play different semantic roles. So from the fact that "rabbit" and "rabbit stage" are not coextensive, it follows that their mental counterparts are distinct. The correct translation of "gavagai" may then be "rabbit" or "rabbit stage" or neither; but it cannot be both. For terms are intertranslatable only if they answer to the same mental representation; otherwise their meanings diverge. If this is correct, translation is determinate; for a fact of the matter determines whether two expressions correspond to the same mental representation. Any residual uncertainty about "gavagai" stems from our ignorance, not from a deficit in semantic facts.

The mental representations cognitive science hypothesizes are inaccessible to introspection. So we have no direct evidence of them. Nevertheless, theorists maintain, we should acknowledge their reality because they are central to a fruitful, powerful, explanatory theory. We should recognize the facts cognitive science appeals to for the same reason we recognize the recondite facts of physics—because they are sanctioned by successful science.

Cognitive scientific theories, like physical theories, are underdetermined by evidence. So perhaps we cannot say whether "rabbit" or "rabbit stage" correctly translates "gavagai". Still, our problem is merely epistemic. If the facts of cognitive function settle the issue, then whether we can ascertain it or not, the meaning of "gavagai" is as determinate as any fact of physics.

But Putnam's thought experiments tell equally against cognitive science's solution. I cannot distinguish beeches from elms; so in my idiolect "beech" and "elm" perform the same function. If linguistic function is determined by cognitive function, the mental counterparts of my terms must be functionally equivalent as well. Since mental symbols are individuated by their function, functionally equivalent mental symbols are identical. But I resolutely deny that "beech" and "elm" mean the same thing. I am convinced that their meanings differ, even though I have no idea what the difference is. Nothing in my mind then determines what I mean by "beech" or by "elm". Moreover, since my twin's ignorance aligns with my own, we remain psychologically indistinguishable. Nevertheless, she does not mean by "beech" what I do. Whether the contents of the mind are determined by self-scrutiny or by science, Putnam's fantasy demonstrates that meanings are not in the head.

In his recent critique of realism, Putnam extends the insight behind his Twin Earth examples.[11] In "The Meaning of 'Meaning'" he showed that nothing in the mind determines meaning; his later work demonstrates that nothing outside it does either. It follows from the Löwenheim-Skolem Theorem (LST) that the axioms of a first-order theory have multiple models in a given domain. Number theory, for example, has numerous set-theoretical models. Nothing in the theory, its models, or the relation of theory to model differentiates the "standard" or "intended" model from "nonstandard" or "unintended" ones; in each model, the formal properties and relations among the natural numbers are preserved. So logic is powerless to identify *the* set-theoretical equivalents of the natural numbers. Each model supplies *a* system of equivalents; but logic provides no basis for favoring one over any of the others.

Putnam's critique rests on the recognition that the LST is indifferent to considerations of content. Any first-order theory, regardless of subject matter, admits of multiple models in a given universe. So a language, being formalizable as a first-order theory, has numerous models in the world. And although the compliants of a term under each interpretation are determinate, selection of any particular interpretation as the bearer of reference is formally indeterminate.

Must multiplicity of models threaten univocality? If meaning and reference are settled by intended interpretation, and if it is determinate that some one interpretation is intended, the existence of unintended models may be a matter of indifference.

But intention is a mental state. A speaker can intend no more than she has in mind. And Putnam's original thought experiments demonstrate that disparate interpretations answer to the contents of a mind. So a speaker's intention cannot determine a unique interpretation of her words. If, for example, she has no idea that H_2O differs from XYZ, she hasn't got what it takes

[11] Hilary Putnam, "Models and Reality," in *Realism and Reason* (Cambridge: Cambridge University Press, 1983), 1–25.

to intend one rather than the other by "water". Both referents accord equally with what she has in mind. There is then no such thing as *the* intended interpretation of a speaker's words. Numerous interpretations conform to her intention. So even if we allow intention to constrain interpretation, indeterminacy is not eliminated.

What should we make of the multiplicity of models? Ordinarily, we invoke model theory to explicate the relation between a syntax and a field of reference. If the formalization of a language is just a regimentation of its syntax, the resulting indeterminacy is innocuous. That syntax does not determine semantics is hardly worth worrying about. But the LST's indifference to content means that Putnam's argument applies even when we construe language more broadly. A comprehensive theory of the language *as used*—incorporating patterns of utterance and inference, dispositions to verbal behavior, even statements of the "plain facts" Searle appeals to and the "semantic facts" favored by cognitive science—admits of multiple models. Enriching our conception of language amounts to adding axioms to our theory. Its models become correspondingly more complex. But broad theories as well as narrow ones, complicated structures as well as simple ones, admit of multiple models. So we cannot elude indeterminacy simply by appealing to a more robust conception of language.

Still, an interpretation that is formally indeterminate is not necessarily wholly indeterminate. Our inability to fix a unique interpretation of number theory may derive from the austerity of our means. Mathematics restricts itself to formal methods; and formal methods are inadequate to the task. But interpretation of a language is bound by no such restrictions. Perhaps we can rectify things by introducing extratheoretical constraints that determine a direct empirical or metaphysical connection between language and its objects. If so, the additional facts that secure reference are facts about the speaker's relation to the world.

The empiricist hope lies in subjecting interpretation to operational as well as theoretical constraints.[12] If the correct interpreta-

[12] Ibid., 8–9, 15–17.

tion is the one that links terms with their compliants, evidence about those compliants should favor that interpretation. Then interpretations that do not answer to the evidence can safely be excluded. Our evidence consists in correlations between particular utterances (our so-called observation sentences) and observed events. Once we adopt such a constraint, interpretations that fail to correlate these utterances with the corresponding events are untenable.

But even if the requisite correlations can be established, they cannot serve as the conduit for reference. For statements of the correlations, like other sentences, are open to multiple interpretations. Moreover, many available correlations connect observation sentences with observationally indistinguishable events. We cannot tell whether "Gavagai!" signals rabbits or rabbit stages just by looking, since there is no discernible difference between rabbits and rabbit stages. Observation then cannot settle the reference of "gavagai".

This is not to say that operational constraints on interpretation are undesirable. It is surely reasonable to require admissible interpretations to correlate observation sentences with observable events. But since those events cannot be individuated apart from language, such constraints cannot be construed as genuinely extratheoretical. They do not escape the indeterminacy that derives from the LST.

Nor can causality serve as the conduit.[13] Unless "cause" has a determinate reference, a sentence like "Rabbits cause us to have a word for rabbits" cannot fix the reference of "rabbit". But "cause", like every other word, admits of multiple interpretations. Along with the rest of our claims, causal judgments map onto the world in numerous ways, yielding divergent models that satisfy our operational and theoretical constraints. Causal statements are part of the structure that requires interpretation, not mechanisms that supply interpretation.

Causal theorists might reply that this response misses their point: the causal theory of reference contends that causality determines reference, not that causal discourse does. Putnam shows, at

[13] Ibid., 17–18.

most, that evidence gleaned from causal discourse does not suffice to identify causality. Nevertheless, causality is a determinate metaphysical relation, and the model bearing that relation to our language determines what our words refer to. The reference of "gavagai" is then determined by a causal fact of the matter, even if that fact is woefully underdetermined by our evidence. Our inability to discover certain facts does not show that those facts do not obtain.

According to a causal theory then, encounters with rabbits rather than rabbit stages (or with rabbit stages rather than rabbits) cause the natives to have the word "gavagai" in their vocabulary. If so, rabbits must affect speakers in ways that rabbit stages do not. It's hard to see what difference there could be. So it's hard to credit rabbits with causal powers that rabbit stages lack.

The difficulty is not just epistemic. For the attempt to divorce causality from causal discourse is doomed. Causality is whatever relation the verb "cause" refers to. So long as the reference of "cause" is indeterminate, so is the relation of causality. If the causal theory of reference is true, then under one admissible interpretation, rabbits in the vicinity cause the natives to have a word for rabbits; under another, rabbit stages in the vicinity cause them to have a word for rabbit stages. Under the first then, "gavagai" refers to rabbits; under the second, it refers to rabbit stages. The reference of "cause", like that of "gavagai" and "refers", is determinate within an interpretation, indeterminate apart from one. Since the causal theory of reference cannot secure an independent, univocal interpretation of "cause", it affords no escape from indeterminacy. The world then does not determine the reference of natural-kind (or any other) terms.

If terms with the same meaning are coextensive, indeterminacy of translation follows directly. If the extension of "gavagai" is indeterminate, that "rabbit" does nor does not have the same extension as "gavagai" is equally indeterminate. There is no fact of the matter.

But we need not detour through Putnam's critique of realism or accept the thesis that intension determines extension to arrive at this result. Although he does not explicitly acknowledge it,

Putnam's argument applies directly to relations between languages. A translation manual is a function modeling one language in another. Every language admits multiple models in any other. And neither the models themselves nor their relations to the original language or to one another determine which of them preserves meaning. Each English model of the native language reflects native patterns of utterance and inference.

> Smerd gavagai, drok gridnip

is reflected in one model as

> If something is a rabbit, that thing is an animal

and in another as

> If some stage is a rabbit stage, that stage is an animal stage.

Each correlates native observation sentences with stimulus-synonymous English ones. In the first,

> Gavagai!

is correlated with

> Rabbit!

In the second, with

> Rabbit stage!

In each, native statements of introspectively plain fact are mapped onto English statements of equally plain fact.

> Kroq *"gavagai"* ygrup *"gavagai"*

maps onto

By *"rabbit"* I mean "rabbit"

in the first; and onto

By *"rabbit stage"* I mean "rabbit stage"

in the second. And so on. Since all satisfy our criteria for an adequate translation, there is no basis for saying that exactly one among these models captures *the* meaning of the native tongue. Each of the models can provide a translation manual. But the translations supplied by the various manuals are incompatible. Translation is indeterminate.

It follows that the introduction of a "language of thought" or a "deep structure", far from alleviating the problem of radical translation, simply provides another instance of it. For the prelinguistic child's task is then the same as the field linguist's. Each seeks to map initially alien utterances onto a language he already has—the linguist, onto his home language; the child, onto his innate language of thought. And as we have seen, the same evidence is available to both. So the child, endowed with a language of thought, is no better off than the linguist. A spoken language admits of multiple models in such a psychological structure, each with an equal claim to be the determinant of meaning.[14] Moreover, none can be singled out as causally responsible for the generation of surface locutions, since "cause" is not univocal. The term "gavagai" no more has a unique mental counterpart than it has a unique English one. Translation, whether into the language of thought or via the language of thought from one spoken language to another, remains indeterminate.

Linguistic competence is not the ability to articulate antecedently determinate ideas, intensions, or meanings; nor is it the ability to reproduce the world in words. We have no such abili-

---

[14] Fodor (*The Language of Thought*, 73) recognizes that translation from a programming language to machine language is indeterminate. But despite his computational model of mind, he does not seem to appreciate the implications of this for human languages. See also his "Fodor's Guide to Mental Representation," *Mind* 94 (1985): 96.

ties. It consists, rather, in mastery of a complex social practice, an acquired capacity to conform to the mores of a linguistic community. It is neither more nor less than good linguistic behavior.

The additional facts that Quine's critics credit derive their status as fact from an interpretive scheme or translation manual.[15] Being products of interpretation, they cannot supply independent constraints on interpretation. Facts they may be; but not facts of the matter that concern philosophy of language.

[15] See Chapter 11.

CHAPTER **5**

Restoration and Work Identity

Things change. Through some changes they retain their identities; through others, they do not. A critical question then is what alterations an object can undergo and remain the same thing. The answer obviously depends on what sort of thing it is. My car remains the same car when the fender is replaced; it is no longer the same car when taken apart and sold for scrap. The university retains its identity as students come and go. The sophomore class changes its identity from one year to the next.

Gradual change with replacement is particularly vexing, as the ship of Theseus shows. Being an ancient vessel, it was made of wood. Over time, as its planks rotted, they were replaced, one by one, until none of the original planks remained. Was it still the same ship? If not, when precisely did it become a new one?

Restoration of works of art threatens to pose the same problem. Suppose restorers replace the disintegrating threads of a tapestry one by one until none of the original threads remains. Is it still the same tapestry? If not, when did its identity change? Or suppose they replace the paint on a canvas as it chips off, one square millimeter at a time, until none of the original paint is left. Is it still the same painting? If not, when was its original identity lost?

These are not idle theoretical questions. For the legitimacy of restoration evidently turns on their answers. Botched restoration ruins works of art. The worry here is that impeccable restoration is equally destructive. Restoration would not only be inaptly named but inordinately difficult to justify if it inevitably wrecked the works it set out to repair.

Supplementation and deletion pose similar problems. But the problems seem more salient where supplementation occurs, so I shall concentrate on these. It should be obvious, though, that the same problems arise whether one is adding to or removing from a work of art.

When a restorer retouches a painting, she introduces onto the canvas something that was not originally there. What results, theorists charge, is a hybrid—in part the work of the original artist, in part the work of the restorer.[1] Since the original was no hybrid, the restored work is not the same painting. The original no longer exists.

This argument rests on the unspoken and, so far as I can tell, unexamined assumption that new paint, once applied, becomes part of the work. Let us call this the integralist assumption. Adherents of the integralist assumption typically do not want to prohibit restoration entirely. They have no qualms about cleaning, for example, for cleaning neither adds anything to, nor removes anything from the work itself. Nor do they object to strengthening or replacing a work's supports. What happens to the back of a canvas is, they assume, a matter of aesthetic indifference. What they object to is any interference with the work's aesthetic elements.

The distinction they rely on is more easily articulated than applied. Whether a given feature plays an aesthetic or a supportive role is not always obvious. Adherents of the integralist assumption sanction removal of varnish only if it functions solely to protect. If it is integral to the work, the varnish is not to be removed. Where we know that the varnish was applied only long after the work was completed, we might be safe in

[1] See Mark Sagoff, "On Restoring and Reproducing Art," *Journal of Philosophy* 75 (1978): 453–469; S. J. Wilsmore, "What Justifies Restoration," *Philosophical Quarterly* 38 (1988): 56–67.

taking its role to be purely protective. But if the original artist applied the varnish or instructed others to apply it, its status is incurably uncertain. It might have some aesthetic role, or the artist might simply have wanted to protect his work. Having no objection to replacing supports, an adherent of the integralist assumption would permit the transfer of a painting from a severely worm-eaten panel—assuming, of course, that the panel was just a support. But the painting may derive subtle aesthetic characteristics from the panel—for example, from the grain and texture of the wood.[2] In that case, the panel is not to be touched.

What is permitted and what prohibited then has to be decided on a case-by-case basis. And verdicts about individual cases may be essentially contestable. Adherents of the integralist assumption condemn some types of restoration out of hand, but they cannot wholly exonerate any.

The prohibition against interference with aesthetic features dooms certain works of art—for example, the painting whose worm-eaten panel is integral to it. Even though disintegration is imminent, replacement of the panel is precluded on pain of destroying the work. Similarly, removal of varnish deemed integral to a painting is impermissible even when the varnish has darkened to the point of opacity. The adherent of the integralist assumption would claim that he is just facing facts. The works I have described, he insists, are doomed. Restorers who pretend otherwise deceive themselves and perpetrate a fraud on the viewing public. Is he right?

Upon installation, a new fender becomes part of an old car. When the fender is dented, the car sustains damage. And damage to the new fender is as regrettable as equally serious damage to any other part of the vehicle. If the integralist assumption is correct, we should view restorers' contributions in the same light, for then the new paint becomes part of the old painting. But we do not. Damage to the inpainting on a restored Rembrandt would hardly be considered as regrettable

[2] Cesare Brandi, "Restoration and Conservation," *Encyclopedia of World Art* (New York: McGraw-Hill, 1966), 12:180.

as equally serious damage to the paint Rembrandt applied. And if vandals defaced the inpainting but left the original paint intact, we would count it lucky indeed. This difference in attitude should at least give adherents of the integralist assumption pause.

So should the practices and canons of behavior of contemporary restorers. Responsible restorers do not treat their own or their colleagues' contributions as components of the works they repair. Nor do they cover their tracks, as perpetrators of frauds are wont to do. Restorers are extremely reluctant to alter authentic elements of a work of art, but have no qualms about removing the effects of previous restorations. This is why debate rages so furiously over which elements are authentic. If the integralist assumption were correct, there would be nothing to debate. For everything on the canvas would be authentic, even the paint applied last week. Moreover, in making repairs, contemporary restorers take pains to ensure that their contributions can be identified and, if necessary, removed by their successors. Inpainting is supposed to be unobtrusive, not indiscriminable.

The restorer's goal is to preserve and repair the work that comes into her hands. If her predecessors' interpolations had become part of that work, she would be professionally committed to their preservation. She would be justified in removing them only in circumstances that would justify removing handiwork of the original artist. The double standard governing current practice would then be untenable. And if her own additions were destined to become part of the work, she would be required to make them as permanent as possible in order to minimize future deterioration. Easy removal would hardly be a desideratum.

Restorers are professionally self-effacing. It is a criterion of successful restoration that their contributions recede from view. Collaborators are not so shy. Even if Gilbert and Sullivan wanted *The Mikado* to look like a seamless whole, neither wanted it to look like a whole to which he contributed nothing. Restorers evidently do not think of themselves as collaborating in the creation of hybrid works.

Neither the practice nor the self-image of restorers conforms to the integralist assumption. Self-deception is not uncommon. So the integralist assumption's failure to mesh with the profession's characterization of itself is, perhaps, not a serious problem. Mismatch with practice is more troubling. For the practice in question is no mere constellation of unthinking behaviors. It is a self-conscious activity guided by a code of conduct that is the fruit of long, careful reflection on the ends and means of the restorer's craft.

What's the alternative? If restorers' interpolations are not integral parts of the works they repair, what are they? I suggest that they are prostheses. Their analog is not a new fender, but an artificial limb. A prosthesis is an artifact that substitutes for an appendage or organ to recover some lost human function. Even though an artificial leg takes the place of a natural one, no one is inclined to consider it part of the body it supports. Nor would any sane person advise an amputee to forego a prosthesis on the ground that in donning an artificial leg he starts down the slippery slope that will eventuate in his being replaced by a robot— a fully mechanical device that looks and acts exactly like him. There may be good arguments against getting a prosthesis. This is not one of them.

No more should we consider restorers' interpolations parts of the works that contain them. Rather, they should be seen as artificial substitutes designed to reenable damaged works to perform their aesthetic functions. A restorer is in effect a physician. Her goal is to alleviate aesthetic disabilities.

Mark Sagoff, one of the most articulate adherents of the integralist assumption, uses "prosthesis" as a term of disparagement.[3] But he considers prostheses decorative rather than functional devices. So he concludes that by resorting to prostheses, restorers transform works of art into mere decorations. Sagoff, however, misunderstands the role of prosthesis. One can walk on an artificial leg, grasp with an artificial hand, chew with false teeth—not, to be sure, as well or as easily as with healthy natural appendages, but that, sadly, is not an option. The patient's choice,

---

[3] Sagoff, "On Restoring and Reproducing Art," 459.

at best, is between a dysfunctional natural organ and a functional artificial one.

Sometimes, to be sure, a prosthesis is used solely to present the appearance of an absent member. Medical science has not yet, for example, developed an artificial eye that enables a person to see. But even here, I think, the purpose should not be construed as mere decoration. Looking disabled can itself be a disability. And looking like an able-bodied person can promote the capacity to act and be treated as one. So even where prostheses only present the appearance of missing body parts, their role is to promote human functioning.

The medical model of restoration vindicates both the self-image and the practice of the profession. Like physicians, restorers remedy where possible, removing impediments to proper functioning. Where a condition is incurable, they equip the disabled to minimize their loss of function. The prosthetic surgeon does not become a parent of the patient whose artificial organ he implants. Nor does the restorer become a creator of the painting whose lacunae she fills.

If the restorer's interpolations are aids to, but not strictly part of the works that embed them, the justification of practice is straightforward. An interpolation should be identifiable so as not to be confused with genuine elements of the work. Our understanding of an artist's originality, style, development, and treatment of a subject would be seriously hampered if the distinction between his handiwork and that of his restorers could not be drawn. Interpolation should be unobtrusive so as not to interfere with the work's aesthetic functions. Clumsy repairs and ostentatious additions draw attention to themselves, distracting and detracting from the art they are supposed to sustain. And interpolations should be easy to remove. History shows that they can fail to reinvigorate the works that embed them, can be unwitting sources of further damage, can be superseded as restoration techniques and aesthetic understanding evolve.

The construal I favor shows that the distinction between aesthetic and supportive additions is spurious. Whether a restorer adds paint to the front of a canvas or glue to the back, her contributions play a supportive role. They help a disabled work perform its aesthetic functions.

It is easy enough, I suppose, to identify human disabilities. But it may be harder to recognize aesthetic ones. A look at some dysfunctional works may help. Consider then the following:

(1) A Dutch gallery picture whose varnish has darkened, making details on the pictorial surface practically impossible to discern;

(2) A Mondrian that vandals have spattered with paint in such a way that its design is impossible to make out;

(3) A fresco whose lacunae are so severe that it is difficult to recognize subject, style, or mood;

(4) A medieval tapestry locked away in a vault because further exposure to light or air would cause it to disintegrate.

What is impeded in all these cases, I contend, is the capacity to perform certain symbolic functions. These functions, which Goodman calls symptoms of the aesthetic, "tend to focus attention on the symbol rather than, or at least along with, what it refers to".[4]

I do not claim (nor, of course, does Goodman) that every functional work of art performs all these functions. But I think that aesthetic disability consists in a work of art's loss or diminution of such of these functions as it once possessed. Thus, I suggest, it is these functions that the restorer seeks to recover. Let us look briefly at them.

Where symbols are *syntactically dense*, minute differences affect identity. Because darkened varnish obscures the painted surface, the viewer of the Dutch painting cannot discern these differences, hence cannot determine what symbol she sees.

Where symbols are *semantically dense*, the finest differences in certain respects affect their reference. Gaps in the fresco make it impossible to tell whether a facial expression is a smile or a sneer, thus impossible to tell whether the figure represents Jesus or Judas.

[4] Nelson Goodman, *Ways of Worldmaking* (Indianapolis: Hackett, 1978), 67–69.

In *relatively replete* symbols, comparatively many aspects are significant. Exact thickness of lines, precise proportions of volumes, minuscule differences in tonality matter to the Mondrian. If vandals' damage occludes or overshadows these, the viewer will be unable to comprehend what and how the work conveys.

In *exemplifying*, a symbol serves as a sample of features it literally or metaphorically possesses. Where damage interferes with the gallery painting's exemplification of tonality, the fresco's exemplification of spirituality, the Mondrian's exemplification of shape, it prevents or hinders the viewer's effectively focusing on features the works are samples of.

Where a symbol exhibits *multiple and complex reference*, it plays a variety of interacting referential roles. If the varnish has darkened to the point where the viewer cannot see what and how the pictures within the gallery picture depict, its allusions to and commentaries on other works, styles, treatments, and subjects are lost.

The tapestry is a limiting case. It may once have performed all the symbolic functions I mentioned. But locked away, inaccessible to viewers, it now performs none. Moreover, the tapestry shows that a work's disability need not consist in any defect in its symbolizing elements. Deterioration of the underlying fabric is sufficient to disable the work. Strengthening the backing would remove the disability, enabling it again to function aesthetically.

Restorers devote their lives to the care and preservation of art *objects*. Can an account that insists on treating works as symbols make sense of that? It can. For the symbols in question are inseparable from the material objects that realize them.

A syntactically dense symbol is uniquely instantiated. Such a symbol cannot survive the destruction of the object that realizes it. If we care about the symbol then, we have reason to preserve and, if necessary, repair the object in which it resides. Since, moreover, works of art are relatively replete, damage along any of comparatively many dimensions threatens to inhibit symbolic functioning. Since works admit of multiple correct interpretations, we can never safely claim that our enumeration of those dimensions is complete. If every aspect of an art object has the potential to function symbolically, it would make no sense to purport to care about the symbol but not about the object.

To do her job, a restorer needs technical expertise backed by knowledge drawn from a variety of fields—spectroscopy, chemistry, and materials science, to name a few. Manual dexterity worthy of a neurosurgeon and nerves of steel also help. Just to figure out what she is trying to do requires knowledge of the work she is dealing with and its art historical context, as well as critical acuity, and delicacy of discrimination.

It is not always obvious whether a work calls for restoration. Some damage is too minor to matter; some too extensive to repair. And sometimes, although local damage can be undone, the overall effect would be a diminution of aesthetic function. This is the reason critics advise against cleaning the facades of buildings in historic districts. By the time the last one was done, the first would be filthy again. Cleanliness and harmony evidently are not jointly realizable.[5] Moreover, not every change that time brings constitutes damage of any kind. Patina often enhances a work's capacity to function aesthetically by, for example, highlighting its exemplification of features we might otherwise overlook. Were the goal of restoration the removal of the effects of time, such patina should go. Since the goal is the reversal of the ravages of time, it ought to be retained.

The multiplicity and interplay of a work's symbolic functions contribute to the difficulty of the restorer's task. What is wanted, typically, is not just to recover a single function, but to recover several at once, preferably without undermining others. This is not always possible. It may be necessary to forego some symbolic functions to preserve or recover others. Transfer of a painting from the panel it was painted on, for example, sacrifices exemplification of features derived from the grain of the wood. If the panel is sufficiently damaged, the transfer is justified. But the aesthetic cost must be acknowledged. The work originally had symbolic functions that could not be recovered. In such cases, the restorer must decide which realizable combination of functions is

[5] Paul Philippot, "Historic Preservation: Philosophy, Criteria, Guidelines," *Preservation and Conservation: Principles and Practice*, ed. Sharon Timmons (Washington: The Preservation Press, 1976), 374.

best on balance. These difficulties are real, but they are just occupational hazards. They pose no problem in principle. One philosophical worry still looms. Unless it can be satisfactorily resolved, we may yet be forced to adopt the integralist assumption.

The identity of a painting is fixed, Goodman and I contend, by its history of production.[6] A painting is a syntactically dense symbol, and such symbols admit of no replicas. No difference is so small as to be insignificant. Even the closest copy is bound to diverge from the original somewhere. So nothing with a different history can be the same painting. This explains why even the best reproductions are not instances of the paintings they purport to reproduce. It does not, however, tell against restorations. Although every difference *between instances* constitutes a difference between the symbols they instantiate, it does not follow that every change *in an individual instance* constitutes a difference in the symbol it instantiates. Indeed, I would urge, such is not the case. Paintings, and the symbols that constitute them, retain their identities through time and the changes time brings. To think otherwise would be to take paintings to be ephemeral indeed. When colors fade, cracks emerge, lacunae develop, and varnish darkens, a painting becomes neither a different symbol nor a different work. Change, per se, does not affect a symbol's identity.

This may seem implausible. So let us look at a related case. A letter *e* printed in a broken typeface would still be an *e*. But the break could undermine access to it, possibly prompting readers to mistake it for a *c*. Inking in the gap would not transform the *e* into another symbol, but would make the extant symbol easier to read. Likewise, I contend, a chipped Giotto Madonna is the same Giotto Madonna it always was. But the lacunae may impede access to it, possibly leading viewers to mistake it for a St. Anne. Filling in the lacunae would not transform it into a different symbol, but would improve access to the symbol already on the wall.

[6] Nelson Goodman, *Languages of Art* (Indianapolis: Hackett, 1972), 116; Nelson Goodman and Catherine Z. Elgin, *Reconceptions* (Indianapolis: Hackett, 1988), 64–65.

The cases are not entirely parallel, though, and the difference between them may give us pause. Letters of the alphabet are replicable. Syntactically dense symbols are not. We might maintain that inking in the *e* yields a distinct but syntactically equivalent replica—another *e*. Pretty plainly, nothing turns on the answer. But any mark a restorer makes is bound to diverge from the one it replaces. Doesn't this make the restored painting a different symbol, hence a different work from the original? It would, if the restorer's marks became part of the symbol that constitutes the painting. But they do not. Rather, they are further changes the enduring symbol undergoes. Unlike chipping and fading, though, if the restorer's marks are well crafted, they enhance rather than diminish aesthetic function.

Restorers' interpolations, unlike the new planks on the ship of Theseus, are not functional equivalents of a painting's original elements. So there is no possibility that a painting might gradually come to consist entirely of restorers' marks. For restorers' contributions, however extensive, do not become part of the work. *The Night Watch* endures only so long as Rembrandt's handiwork can perform its aesthetic functions. Restorers' contributions may increase its longevity, but they cannot take over its job. When the damage is so great that even with the aid of prostheses, what remains of Rembrandt's handiwork can no longer perform its original aesthetic functions, the work ceases to exist. Since paintings are syntactically dense symbols, nothing by another hand can perform precisely the functions of Rembrandt's *The Night Watch*. The loss is irrecoverable.

It is possible, of course, that we might mistake a collocation of restorers' marks for *The Night Watch*, just as it is possible that we might mistake a reproduction or a forgery for the work. Neither error would transform the object of our attentions into *The Night Watch*. Both would make our attributions wrong.

Whether a badly damaged painting has been destroyed may be a matter of dispute. The painting exists only so long as its aesthetic functions can be performed. But the point of no return is not easy to locate, and knowledgeable critics may differ over its exact location. There are several legitimate grounds for disagreement. The functions I have characterized as aesthetic are extraordinarily fine-grained. So opinions may diverge as to exactly what functions a particular work performs. Damage that would destroy a work's capacity to exemplify some precise measure of melancholy might not affect its capacity to exemplify sadness. Differences over which of these features the painting exemplified would give rise to divergent assessments of the seriousness of the damage. Moreover, there need be no unanimity about a damaged work's prospects. Some may believe restoration can recover an acceptable level of functioning, others believe not. Finally, it is obvious that works can survive some loss of aesthetic functioning. But it is not obvious how much. So critics who agree about what the work's original aesthetic functions were, which ones were lost, and what restoration can recover might still disagree about whether the work survives the loss.

In works of art, the finest details carry enormous weight. The boundary between significant and insignificant losses cannot be precisely delineated. So indecision about borderline cases is inevitable. But from such indecision nothing directly follows about how borderline cases should be treated. Perhaps restorers should undertake heroic measures in a desperate effort to resuscitate. Perhaps they should direct their energies elsewhere, since their efforts to save such works are apt to be futile.

I doubt that the issue admits of a general resolution. It is likely to depend on the particulars of each case: the importance of the work in question, the level of aesthetic functioning one can plausibly hope to restore, the alternative uses of available time, talent, and other resources, and so on. Only by carefully weighing these, I think, can a responsible decision be reached.

Although the situation is more desperate here, the questions are the same as those that arise in attempting to decide whether any other work ought to be restored. In these cases, however, the restorer at least has the consolation of knowing that her ministrations will not leave the work worse off than it was before.

A work of art, I have urged, performs a variety of quite specific symbolic functions. When damage interferes with its performance of those functions, it becomes disabled. The restorer's goal is to enable the work to overcome its disabilities, either by removing impediments or implanting prostheses. If the work largely recovers its capacity to perform its symbolic functions, restoration is successful. The work endures.

CHAPTER 6

Translucent Belief

One of the more perplexing problems in the philosophy of language involves the interpretation of belief ascriptions—sentences of the form

> *R* believes that *p*

where the expression that replaces *p* (the content of the ascription) has the syntactical form of a sentence. The problem is to say how such contents are to be interpreted.

Philosophical lore has it that our alternatives are exactly two: either an ascription of belief is transparent or it is opaque.[1] If it is transparent, the content is just the sentence it appears to be, and its interpretation is straightforward. If it is opaque, this is not the case. Philosophers disagree about how opaque constructions are to be interpreted, but this much is generally agreed: expressions occurring in opaque contexts do not have the same referents as they have in transparent contexts; and the wording of an opaque construction affects the truth value of the sentence that contains it in ways in which the wording of a transparent construction does

[1] Some may be one, some the other. Indeed, some tokens of a single type may be transparent, others opaque. But tradition has it that unless a token is transparent, it is opaque. There is no third alternative.

not. In a transparent context we, as it were, look through our words to their referents; in an opaque context we attend to the words themselves.

The conviction that the disjunction "transparent or opaque" exhausts our options strikes me as mistaken. For many ascriptions of belief, neither construal is satisfactory. An intermediate position is wanted—one in which due attention is paid both to the words that are the medium of reference and to the things that are the objects of reference. I suggest then that we extend the metaphor and say that ascriptions of belief are typically neither transparent nor opaque, but *translucent*. A medium is literally translucent if it transmits light, but diffuses it in such a way that the outlines of objects cannot be clearly discerned. Correspondingly, a linguistic construction is metaphorically translucent if it transmits reference, but in such a way that the limits on paraphrase are not obvious. Just as we must attend to the nature of a literally translucent medium to understand what can be seen through it and what distortions and obscurities are to be expected, we must do likewise in the metaphorical case to understand how reference is transmitted, distorted, and obscured in ascriptions of belief. Just as literally translucent media range from the all but transparent to the all but opaque, so do metaphorically translucent media. And just as a translucent medium may block some light waves and transmit others, so a metaphorically translucent medium may block some substitutions but permit others. We should expect the limits on permissible paraphrase in translucent contexts to vary considerably—some such contexts being more nearly transparent, others more nearly opaque.

My strategy in this paper is as follows: I review familiar accounts of the interpretation of belief ascriptions to identify their characteristic strengths and weaknesses. I then formulate an account in which such ascriptions are construed as translucent. I suggest that it retains important virtues of its predecessors and avoids their major failings. With suitable modifications, my account can be extended to cover ascriptions of other so-called "propositional attitudes"—fears, hopes, expectations, and the rest. But I will not attempt that project here. I should emphasize that it is no part of my project to say what beliefs are. I am concerned exclusively with the interpretation of expressions that follow the phrase "believes that" in ascriptions of belief. I think significant progress can be made on that problem without settling the reference of the term "belief"—without, that is, solving substantive problems in the philosophy of mind.

When belief ascriptions are construed transparently, the terms in their content statements have their normal referents.[2] This allows for an easy understanding of some of the ways beliefs are said to function in epistemology. We classify beliefs as true or false, as confirmed, disconfirmed, or unconfirmed, as rational or irrational, and so on. The belief ascribed to Sam in

(A)   Sam believes that kangaroos are carnivores.

is true only if kangaroos eat meat; it is confirmed only if there is evidence that they do so; it is rational only if there is sufficiently good reason to think that they do so; and so forth. Information about the dietary habits of kangaroos is plainly germane to the epistemic status of Sam's belief if in the statement of that belief we make reference to kangaroos. But if in the context of belief ascriptions "kangaroo" does not refer to kangaroos, it is hard to see why or how such animals should matter to the epistemic evaluation of Sam's belief. It is then a virtue of the transparent construal that it preserves a clear connection between the interpretation of belief ascriptions and the epistemic status of the beliefs they ascribe.

In transparent contexts, all coextensive terms are intersubstitutable *salva veritate*. So it follows from (A) that

(B)   Sam believes that marsupials of the family macropodidae are carnivores.

even if Sam, being ignorant of taxonomy, is in no position to affirm (B). Indeed, for every belief a person ascribes to himself, indefinitely many other ascriptions are equally applicable—namely, all those which result from the substitution of coexten-

---

[2] Or at least their occurring in the context of a belief ascription does not affect their reference. Metaphor, ambiguity, vagueness, and deviant usage may occur in belief ascriptions just as in any other sentences.

sive terms for the terms in his ascription. Many of these he will not recognize as statements of his belief; indeed, many he will not even understand. Moreover, all such paraphrases are on a par. Those whose truth he acknowledges have no special standing.

Although Sam is prepared to say that kangaroos are carnivores, he is under the misapprehension that carnivores are eaten as meat, not eaters of meat. As regards the eating habits of kangaroos,

(C)   Sam believes that kangaroos are not meat eaters.

His willingness to accept the contents of both (A) and (C) shows Sam to have incompatible beliefs about the place of kangaroos in the food chain. Since he is unlikely to be in a position to feed or to be fed to kangaroos, this seems but a minor defect in his doxastic system.

One of the strengths of a transparent construal is that it enables us to ascribe mutually incompatible beliefs to an agent. Although we often do so, this is apt to be overlooked in the formulation of theories of belief ascription. The conviction that it is always a mistake to harbor incompatible beliefs slides easily over into the conviction that it is always a mistake to ascribe them.[3] But since we regularly ascribe such beliefs, our system should have the resources to say so. And since people actually harbor such beliefs, it should not follow from our system that such ascriptions are inevitably in error. In these respects, the transparent construal of belief ascriptions appears satisfactory.

But a difficulty remains. On a transparent reading, (A) is equivalent to

(D)   Sam believes that kangaroos are meat eaters.

So Sam's beliefs are not merely incompatible, they explicitly contradict each other. What looked to be a minor flaw in his factual beliefs now looks to be a serious logical failing.

[3] Quine's "principle of charity" encourages this tendency. See W. V. Quine, *Word and Object* (Cambridge: MIT Press, 1960), 59.

Still, it seems wrong to convict Sam of an error in logic. He does not, after all, recognize the contents of both (C) and (D) as statements he believes. He is well aware that one of them must be wrong, even if he has no idea which one. Moreover, instruction in logic will not avail him in recognizing and correcting his error. But instruction regarding the eating habits of kangaroos and carnivores, or regarding the interpretation of the terms "kangaroo" and "carnivore" might do so.[4] If, however, our ascriptions are transparent, we have no basis for calling his error zoological or semantic rather than logical. Epistemology requires a more sensitive device for delineating beliefs than the transparent reading provides.

Do opaque construals fare any better? In considering the matter we should distinguish between Quinean and Fregean accounts. The former take the content of a belief ascription to be semantically simple; the latter take it to be semantically complex.

Thus Quine construes the phrase "believes that kangaroos are carnivores" as an unbreakable one-place predicate that applies to Sam if (A) is true. "Kangaroo" then no more occurs as a term in (A) than "cat" occurs as a term in "cattle".[5] This enables us to say that sentences (A)–(D) ascribe distinct, logically independent beliefs to Sam. But Quine's criterion is too sensitive in that any syntactic difference in the content of belief ascriptions marks a difference in the beliefs ascribed. And it is too insensitive in that syntactically identical sequences with different interpretations count as the same belief.[6] Attempts to evade the latter difficulty by relativizing ascriptions to a language or to someone's use of a language reintroduce the problematic semantic features that the account was designed to avoid.[7] Moreover, if belief contents are construed as uninterpreted syntactical sequences, it is not clear that they have any bearing on epistemology. Questions about the

[4] Not that it is always wrong to ascribe explicitly contradictory beliefs to someone. But such an ascription does not seem to capture the nature of Sam's error.
[5] Quine, *Word and Object*, 211–216.
[6] See Alonzo Church, "On Carnap's Analysis of Statements of Assertion and Belief," *Analysis* 10 (1950): 97–99.
[7] Quine, of course, recognizes this and concludes that the vocabulary of propositional attitudes will have no role in limning the true and ultimate structure of reality. See *Word and Object*, 221.

truth of, or evidence for, beliefs so construed receive no satisfactory answer.

Perhaps what is wanted is an intermediate position—one in which such expressions are credited with interpretations distinct from but systematically related to the interpretations that they have in transparent contexts.[8] This is the sort of account that Frege seeks to provide. He contends that the referent of a term in an opaque context is the sense that term has in a transparent context.[9] (A) then contains no terms that refer to kangaroos or carnivores. But it contains terms that refer to the senses of "kangaroos" and "carnivores". And since the sense of a term determines its reference, (A) is still related, albeit indirectly, to such animals.

In an opaque context, only terms with the same sense are intersubstitutable. If "kangaroo" and "marsupial of the family macropodidae" differ in sense, (B) cannot be obtained from (A) by substitution. In that case (A) and (B) ascribe distinct beliefs to Sam.

Problems remain. Since it is not clear which expressions (if any) have the same sense, it is not clear which expressions are intersubstitutable in opaque contexts. Moreover, even if two expressions are in fact alike in sense, someone might mistakenly think otherwise. For, like the rest of our beliefs, our semantic beliefs are subject to error. If "carnivore" and "meat eater" have the same sense, (A) is equivalent to (D), whether Sam thinks so or not. Sam still stands convicted of harboring logically incompatible beliefs—those ascribed in (C) and (D)—even though he does not take himself to believe that kangaroos are meat eaters, and hence has no reason to think that he has any belief that conflicts with (C).

We might avoid this outcome by relativizing our interpretations to idiolects.[10] Since "carnivore" and "meat eater" have dif-

---

[8] See David Kaplan, "Quantifying In," in *Words and Objections*, ed. Donald Davidson and Jaakko Hintikka (Dordrecht: Reidel, 1969), 206–242. Although I agree with Kaplan about the need for an intermediate position, we disagree about how to satisfy that need.

[9] Gottlob Frege, "On Sense and Reference," *Translations from the Writings of Gottlob Frege*, ed. Peter Geach and Max Black (Oxford: Basil Blackwell, 1970), 59.

[10] Such a move is decidedly un-Fregean, for Frege insists that senses are public, shared entities. See Frege, "On Sense and Reference," 59.

ferent senses in Sam's idiolect, (D) cannot be obtained from (A) by substitution. But upon learning that the terms are commonly taken to have the same sense, Sam is apt not only to revise his idiolect, but also to regard his previous understanding of the terms as mistaken. If, however, senses are determined by idiolects, he has no reason to do so. For in that case his usage is not mistaken, just idiosyncratic. And given the senses the terms have in his idiolect, there is no conflict between (A) and (C). Sam's beliefs are not incompatible; his usage is simply unusual. Relativizing interpretations of opaque constructions to idiolects thus obliterates errors rather than explaining them. We seek an account that explains what is wrong with believing both that kangaroos are carnivores and that they are not meat eaters—not one that denies that there is anything wrong with doing so.

Frege's account is open to another objection. We regularly permit substitutions of terms whose senses plainly differ. For example, Max's state of mind with respect to a single arrest might be described as:

(E)   Max believes that the brains behind the Brinks job has been arrested.

or as:

(F)   Max believes that the Brooklyn bank robber has been arrested.

"The brains behind the Brinks job" and "the Brooklyn bank robber" manifestly differ in sense. What allows for their intersubstitutability in statements of Max's belief is his conviction that they refer to the same individual. But it will not do to construe the context as transparent. For even though the fellow—S. J. Lawless, to give him a name—is also responsible for the Sixth Avenue sapphire snatch, Max does not know it. He considers that crime unsolved and thinks its perpetrator still at large. So although some identifying descriptions of Lawless can be freely substituted for one another in the statement of Max's belief, others cannot. (E) and (F) are statements of Max's belief in a way in which

(G)   Max believes that the Sixth Avenue sapphire snatcher has been arrested.

is not. There are then nontransparent contexts in which sameness of sense is not the criterion for intersubstitutability of terms. Frege's account cannot accommodate these.

Many other theories of the interpretation of belief ascriptions have been proposed. But they are largely variations on themes I have already identified. Since my purpose here is to motivate consideration of an alternative approach, it is perhaps unnecessary to detail the ways in which I take the various neo-Fregean interpretations to come to grief.

I suggest that the interpretation of a belief ascription typically turns on both the classification and the truth conditions of the content sentence. In its dependence on classification, my account is reminiscent of opaque construals; in its dependence on truth conditions, it is reminiscent of transparent ones. It is my contention that by combining aspects of each we obtain a more adequate account than either alone provides.

Following Nelson Goodman and Israel Scheffler,[11] let us introduce the schema *p-sentence* whose replacements denote all and only those sentences which are paraphrases of the sentence that replaces *p*. Very roughly, a *p*-sentence is a sentence to the effect that *p*. A replacement for the schema *p-sentence* then is a general term that denotes sentences. "Kangaroos-are-carnivores-sentence" denotes all and only paraphrases of "Kangaroos are carnivores". And "kangaroos-are-carnivores-sentence" is coextensive with "kangaroos-are-meat-eaters-sentence" if and only if the two terms denote exactly the same sentences.

The introduction of *p*-sentences poses no particular problem.[12]

[11] Nelson Goodman, "On Likeness of Meaning," in *Problems and Projects* (Indianapolis: Hackett, 1972), 221–230; Israel Scheffler, "An Inscriptional Approach to Indirect Quotation," *Analysis* 14 (1954): 83–90. Notice that what Scheffler, Goodman, and I do with *p*-labels is very close to what Sellars does with dot quotes. See Wilfrid Sellars, *Science and Metaphysics* (London: Routledge and Kegan Paul, 1968), 80–90.

[12] It is sometimes argued that the introduction of terms by means of such schemata would require the language to have infinitely many semantically

We regularly recognize sentences as paraphrases of one another. And our judgments of paraphrase can serve as a basis for classifying diverse sentences as instances of a common *p*-sentence. Ordinary standards of paraphrase, of course, do not yield necessary and sufficient conditions for the instantiation of a particular *p*-sentence. But this is a common feature of our understanding of general terms. We are equally without necessary and sufficient conditions for the instantiation of the term "home," but we have no difficulty using the term or recognizing homes when we encounter them. There seems no compelling reason to impose more rigorous standards on the terms we use to classify linguistic items than we impose on those we use to classify nonlinguistic ones.

We can begin our explication of (A) then by saying

(A)   Sam believes that kangaroos are carnivores.

only if

(A′)   Sam cognitively favors some kangaroos-are-carnivores-sentence.

I use the vague and unfamiliar term "cognitively favors" to avoid unwanted associations suggested by more familiar mentalistic terminology. Without defining "cognitively favors", I can say something about its application. Among the sentences an individual cognitively favors are those he is inclined to utter, inscribe, or assent to sincerely. His dispositions to verbal behavior determine these, and the evidence that he favors them derives from the

---

independent primitives. Such a language, it is claimed, would be unlearnable. See Donald Davidson, "Theories of Meaning and Learnable Languages," in *Logic, Methodology, and Philosophy of Science* (Amsterdam: North-Holland, 1966), 383–394. Actually, there is no problem. Even though the replacements of the schemata are one-place predicates, there is no reason to think that they have to be learned one by one. See Robert Schwartz, "Infinite Sets, Unbounded Competences, and Models of Mind," *Perception and Cognition: Minnesota Studies in the Philosophy of Science*, ed. C. W. Savage (Minneapolis: University of Minnesota Press, 1978), 9:183–200; and Nelson Goodman, "Splits and Compounds," in *Of Mind and Other Matters* (Cambridge: Harvard University Press, 1984), 77–80.

sentences he actually employs. Typically, however, his dispositions to verbal behavior do not exhaust the class of sentences he
cognitively favors. If he is self-deceptive, insincere, lacking in self-
awareness or in linguistic competence, or simply reticent, he will
cognitively favor sentences he has no inclination to utter, inscribe,
or assent to. So an individual may cognitively favor sentences
without knowing it. The evidence that he does so comes from his
nonverbal behavior, together with the equivocations, hesitations,
hints, and hedges in the sentences—if any—that he employs. Both
verbal and nonverbal behavior then yield evidence of cognitive
favoring, evidence that is used in our ascriptions of belief.[13]

Sentences are symbols in a language. If the syntactical sequence
that an agent utters, inscribes, or accepts has different interpretations in different languages and if contextual cues are too meager
to decide among the rival interpretations, we cannot tell which
sentence he cognitively favors. Before we can use such a sequence as the basis for a belief ascription, we must settle on an
interpretation.

Ascribing a belief to someone then involves identifying and
classifying a sentence we take him to cognitively favor. That
sentence need not be the one that replaces p in our schema. If Sam
speaks only French, our basis for (A) might be his cognitively
favoring

Les kangourous sont carnivores.

Nor need we know exactly which sentence he cognitively favors.
All we need know is that he cognitively favors *some* sentence that
is correctly paraphrased as "Kangaroos are carnivores".

[13] These features enable the notion of cognitively favoring to function in an
explication that comprehends not just beliefs that the agent can readily articulate, but also beliefs that he does not acknowledge, and even ones that he does
not recognize as his own. Moreover, it allows for the ascription of beliefs to
human infants and to nonhuman animals. An infant or an animal who displays
sufficiently sophisticated differential behavior as regards a particular situation
can be said to cognitively favor a sentence to the effect that that situation obtains.
See W. V. Quine, "Quantifiers and Propositional Attitudes," in *The Ways of
Paradox and Other Essays* (New York: Random House, 1966), 194; and Ruth
Barcan Marcus, "Rationality and Believing the Impossible," *Journal of Philosophy*
80 (1983): 332–333.

A person's cognitively favoring a sentence normally reflects his commitment as to how things stand as regards the entities denoted by its terms. So ascribing a belief typically involves imputing such a commitment. This requires adding a truth condition. The belief ascribed to Sam in (A) is true just in case there are such things as kangaroos and whatever is a kangaroo is a carnivore. That is, just in case:

$$(\exists x)(x \text{ is a kangaroo } \& (y)(y \text{ is a kangaroo} \supset y \text{ is a carnivore}))$$

If we incorporate this into our explication of (A), we get

(A″)   ($\exists s$)(Sam cognitively favors s & s is a kangaroos-are-carnivores-sentence & (s is true $\supset$ (($\exists x$)(x is a kangaroo & (y)(y is a kangaroo $\supset$ y is a carnivore)))))

We cannot represent

Kangaroos are carnivores.

simply as

$(x)(x \text{ is a kangaroo} \supset x \text{ is a carnivore})$

for the latter is true if kangaroos do not exist. And Sam cannot plausibly be said to believe that kangaroos are carnivores without its committing him to the existence of kangaroos and carnivores.

This imputation of ontological commitment might seem to raise difficulties for the interpretation of fictional beliefs, especially if we admit no fictive entities into our ontology. But if we accept Goodman's account of fictive language,[14] there is no real problem. According to Goodman, what matters in the interpretation of such language is not what fictive terms denote, but rather what terms denote them. So the fictional sentence

[14] Nelson Goodman, *Languages of Art* (Indianapolis: Hackett, 1976), 21–27. See also Catherine Z. Elgin *With Reference to Reference* (Indianapolis: Hackett, 1983), 43–50.

Leprechauns are wee people.

is explicated as

Leprechaun-descriptions are wee-people-descriptions.

Accordingly,

(H)   Pat (fictively) believes that leprechauns are wee people.

is explicated as

(H′)   ($\exists s$)(Pat cognitively favors s & s is a leprechaun-descriptions-are-wee-people-descriptions-sentence & (s is true $\supset$ ($\exists x$)(x is a leprechaun-description & (y)(y is a leprechaun-description $\supset$ y is a wee-people-description))))

Notice how this differs from the belief of the more gullible Mike.

(I)   Mike believes that leprechauns are wee people.

The content of his belief is a false factual sentence. (I) is explicated as

(I′)   ($\exists s$)(Mike cognitively favors s & s is a leprechauns-are-wee-people-sentence & (s is true $\supset$ ($\exists x$)(x is a leprechaun & (y)(y is a leprechaun $\supset$ y is a wee person))))

The difference in the ontological commitments imputed to Pat and Mike are found in the truth conditions set out in (H′) and (I′) respectively. Pat's belief commits her to the existence of leprechaun-descriptions and wee-people-descriptions, whereas Mike's commits him to the existence of leprechauns and wee people. The descriptions exist, so the truth condition on Pat's belief is satisfied; but leprechauns and wee people do not, so the truth condition on Mike's belief is not.

The truth condition, like the rest of the explication, is strictly extensional. If the predicates have their normal interpretations, the beliefs ascribed to Sam in (A), (B), and (D) have the same truth condition; for "kangaroo" is coextensive with "marsupial of the family macropodidae" and "carnivore" with "meat eater". Sam, of course, need not know this. For nothing in the explication requires him to be cognizant of the truth conditions of the sentences he cognitively favors.

Indiscriminate use of the truth predicate produces paradoxes. But they can be avoided by incorporating restrictions into the truth theory for a language.[15] Such restrictions may require belief ascriptions to be construed as metalinguistic. And the truth conditions on certain content sentences will be more complicated than my examples suggest. Just what complexities are involved depends on which truth theory is used to interpret the language in which the ascription occurs.

The truth condition yields the virtues of the transparent reading; the classification, those of the opaque. The features required for "belief" to perform its epistemological function follow from the truth condition. A belief is true just in case its truth condition is satisfied; it is warranted to the extent that there is evidence that its truth condition is satisfied; and so on. The truth condition is what relates (A) to kangaroos and carnivores in such a way that the truth value of the belief ascribed to Sam depends on those animals, and its warrant on what is known about them.

Moreover, the truth condition enables us to explain what is involved in harboring incompatible beliefs and why incompatibilities among beliefs often go unnoticed. Beliefs are incompatible if their truth conditions are not simultaneously satisfiable. Since the agent need not know the truth conditions on his beliefs, he may be in no position to tell whether this is the case. Indeed, sentences that are alike in truth conditions are often classified as disparate *p*-sentences and on the basis of criteria that are more restrictive than coextensiveness of corresponding terms. As a result, the logical relations among the several sentences an agent cognitively favors are apt to be obscured by their classification. So

[15] See Alfred Tarski, "The Concept of Truth in Formalized Languages," in *Logic, Semantics, and Metamathematics* (Indianapolis: Hackett, 1983), 152–278; and Saul Kripke, "Outline of a Theory of Truth," *Journal of Philosophy* 72 (1975): 690–716.

even if the agent is aware that he cognitively favors certain sentences, he may have no way to tell whether they conflict.

The sentences we classify as instances of a common *p*-sentence are paraphrases of one another. Thus, those which paraphrase "Kangaroos are carnivores" are all kangaroos-are-carnivores-sentences. The difficulty is to settle on a criterion for paraphrase.

It is tempting, perhaps, to construe paraphrase in terms of synonymy. Then sentences are paraphrases of one another just in case their corresponding terms are synonymous. But we have no clear criterion for synonymy. It is not obvious whether, for example, "meat eater" is (close enough to) synonymous with "carnivore" for the content of (D) to count as a paraphrase of that of (A). Moreover, there are cases in which likeness of meaning, however construed, plays little part in determining the extension of our *p*-sentences. However it is that we decide whether the content of (F) is a paraphrase of that of (E), and so a Lawless-has-been-arrested-sentence, it is plainly not on the basis of anything like synonymy of their corresponding terms.

A related problem is seen in Saul Kripke's puzzle about belief.[16] Pierre, a monolingual French child, sincerely asserts

(J)   Londres est joli.

on the basis, perhaps, of stories he has heard and pictures he has seen. During a war, he is evacuated to a particularly dismal section of London. Since his new acquaintances speak no French, he learns to speak English directly, as a native speaker does. He comes then to assert

(K)   London is not pretty.

But being unaware that "Londres" and "London" denote the same city, he does not withdraw his previous assertion of (J). What beliefs can we reasonably ascribe to Pierre?

Plainly the beliefs ascribed in

[16] Saul Kripke, "A Puzzle About Belief," in *Meaning and Use*, ed. A. Margalit (Dordrecht: Reidel, 1979), 239–283.

(L)   Pierre believes that London is pretty.

and

(M)   Pierre croit que Londres est joli.

have the same truth condition. So the question whether they ascribe the same belief turns on whether they have the same classification condition. Typically we treat translations of *p* as *p*-sentences. So ordinarily we have no qualms about classifying (J) as a London-is-pretty-sentence, nor about ascribing to someone who cognitively favors (J) the belief that London is pretty. What gives pause about Pierre's case is the fact that the correct literal translation of a sentence he asserts is one he explicitly disavows.

Sentences, like other entities, can be variously classified; and in different contexts or relative to different interests, different classifications may be appropriate. Frequently we ascribe beliefs in order to explain a person's behavior (including his verbal behavior). If such is our goal here, then we should not treat the translation of (J) into English as a *J*-sentence.[17] For a plausible explanation of Pierre's untroubled assertion of sentences that are incompatible is that he is unaware of their incompatibility. This suggests that we explicate his beliefs as follows:

(M′)   (∃*s*)(Pierre cognitively favors *s* & *s* is a Londres-est-joli-sentence
        & (*s* is true ⊃ (∃*x*)(*x* = London & *x* is pretty)))

and

(N′)   (∃*t*)(Pierre cognitively favors *t* & *t* is a London-is-not-pretty-
        sentence & (*t* is true ⊃ (∃*y*)(*y* = London & ~*y* is pretty)))

---

[17] This is not the only reasonable goal. We could take Pierre's situation to show what problems can arise from learning more than one language "like a native." We might then contend that because Pierre learned the languages independently of each other he believes both that London is pretty and that London is not pretty. We would then treat (J) as a London-is-pretty-sentence. Indeed, whatever philosophers think of Kripke's puzzle, it should give the Berlitz people pause.

Pierre's assertion, "Londres est joli", then is not classified as a London-is-pretty-sentence, despite its being a translation of, and having the same truth condition as an English sentence that is so classified. Pierre's beliefs are, on this reading, jointly untenable, but not logically contradictory.

Does this mean that we should entirely abandon the practice of treating translations of p as p-sentences? If we require a general criterion that determines for every p what counts as a p-sentence, we seem forced to do so. This outcome is unwelcome, however; for the practice is useful and rarely gives rise to difficulties. Moreover, without it we have no way to ascribe beliefs that we can understand to people who do not speak our language. Instead, it seems best to deny that we either have or need such a criterion: p-sentences are paraphrases of p, and there is no general criterion for determining the limits on permissible paraphrase. The goal of an explication of belief ascriptions is then not the identification of a single, rigid standard that determines for every R and for every p whether R believes that p.

The quest for such a standard was, in any case, misguided. We seek to explicate actual belief ascriptions, and we are fairly flexible about the beliefs we are prepared to ascribe. We adjust the parameters on permissible paraphrase to suit our several purposes, to answer to our various interests, to achieve our assorted ends. An explication of belief ascriptions should yield a schema that shows what such adjustments involve. It need not yield an algorithm for making them.

It might be thought that I have given up too quickly. Perhaps there are general rules for classifying the sentences we cognitively favor, but more complicated rules than I suggested. If so, the problems I mentioned might just show that the rules have to contain defeasibility conditions or exceptive clauses. For example, all translations of p are p-sentences except when. . . . This suggestion is not promising, however, because the cases that perplex us typically involve mistaken beliefs. And the human capacity to make errors seems far to outstrip our capacity to anticipate them and to build into our rules of interpretation mechanisms to accommodate them as they arise. Unless we can tell in advance just what mistakes will be made, we cannot tell what defeasibility conditions or exceptive clauses need to be incorporated into our rules.

Typically, of course, we have little difficulty deciding how broadly to construe paraphrase, and in what respects paraphrases must agree. So typically we have little difficulty saying which sentences count as paraphrases of each other. But in different contexts and relative to different interests such matters are decided differently.

When we are concerned exclusively with the situation a belief is about, we often take the truth condition to determine the classification condition. Thus, a counterespionage agent interested in getting the goods on Ortcutt might take any sentence s whose truth condition is

$$(s \text{ is true} \supset (\exists x)(x = \text{Ortcutt} \& x \text{ is a spy}))$$

to be an Ortcutt-is-a-spy-sentence, and so take any belief ascription with that truth condition to ascribe the belief that Ortcutt is a spy. This amounts to construing the ascription transparently.

At the opposite extreme, if a sincere avowal is sufficiently bizarre, we might permit nothing but its syntactic replicas to count as paraphrases. If Arlo avers, "People are prime numbers", we might take only replicas of his remark to satisfy the classification condition. In such cases, we are primarily interested in the way a peculiar assertion is to be classified, not in what it would take for that assertion to be true. And because of its peculiarity, only replicas are counted as satisfying the classification condition. But even though we focus on the classification condition, it is important to notice that these belief ascriptions have truth conditions as well. Indeed, it is because we recognize that their truth conditions are those of blatant falsehoods or trivial truths that we retreat to what amounts to an opaque construal.

Ascriptions of belief typically fall between these extremes. They are, in my terminology, translucent. Sensitivity to contextual factors is thus required to identify the appropriate p-sentence and to fix its extension. In the explication of a casual conversation among neighbors, Sam's assertion, "Kangaroos are carnivores", might be classified as a kangaroos-are-meat-eaters-sentence on the grounds that the coextensiveness of "carnivores" and "meat eaters" is widely known. In such a context, however, we should be reluctant to classify it as a marsupials-of-the-family-macripodidae-are-meat-eaters-sentence, for the coextensiveness that classification is based on is not common knowledge. In an explication of a discussion among zoologists, however, his remark might well be so classified. Whether Sam's utterance counts in favor of (B), of (D), of neither, or of both, depends on how it is classified, and its classification depends in turn on a variety of factors including the context in which and the reasons why his assertion is being considered.

The criterion we employ need not take Sam as the final authority. Neither the "common knowledge" that grounds the first classification nor the more specialized knowledge that grounds the second need be knowledge that Sam shares. Sometimes it is reasonable to ascribe to a person beliefs that he does not recognize as his own.

There are, to be sure, cases in which the paraphrases that the agent accepts determine the extension of the p-sentence under which they are to be classified. Our basis for classifying the contents of (E) and (F) as Lawless-has-been-arrested-sentences is Max's conviction that they all refer to Lawless. And our reason for refusing to so classify (G) is his conviction that the Sixth Avenue sapphire snatch was perpetrated by someone else entirely. So even though (E) and (G) have the same truth condition, where Max's views on the matter are authoritative, they have different classification conditions. "Lawless-has-been-arrested-sentence" then does not comprehend every sentence that asserts that Lawless has been arrested. Nor need it comprehend only such sentences. If the Brooklyn bank robberies were in fact perpetrated by someone else, Lawless's capture leaves the Brooklyn bank robber at large. Still, by our criterion, "The Brooklyn bank robber has been arrested" is a Lawless-has-been-arrested-sentence. (F) then has the same classification condition as (E), but a different truth condition.

These features are not peculiar to cases in which it is the agent's convictions that determine the classification condition. Sometimes, as we saw, our classification criterion is "common knowledge"—an amalgam of knowledge, ignorance, and error which is publically shared. Because such "knowledge" falls short of omniscience, there are sentences asserting p that are not p-sentences. And because it falls short of infallibility, there are p-sentences that do not assert that p.

The foregoing account yields no criterion of cognitively favoring. As a result, it provides no basis for ruling that some sentences are a priori unbelievable. There may, of course, be sentences that nobody in fact cognitively favors (sentences with seventeen nested negations, perhaps). But this is a matter for empirical psychology to discover; it is not to be decided in advance. Consequently, we cannot claim that certain things (for instance, contradictions, violations of natural law, sufficiently improbable events) are intrinsically incredible. This strikes me as a welcome result. For if sentences asserting such things are in fact accepted— if, for example, they are sincerely uttered, inscribed, or assented to with every evidence of being comprehended—it is better to construe them as irrational beliefs than to deny that they are beliefs at all.

It follows that certain seemingly legitimate inferences cannot be drawn. If R believes that p, and R believes that q, it does not follow that R believes that p & q. For he may be prepared to sincerely utter, inscribe, or assent to each conjunct separately, without being prepared to sincerely utter, inscribe, or assent to their conjunction. This enables us to make sense of harboring beliefs that contradict one another without thereby harboring any contradictory beliefs. R may cognitively favor p and cognitively favor $\sim p$ without cognitively favoring p & $\sim p$. Indeed, doing so is not uncommon. We are all too often willing to assent to each of a pair of contradictory sentences ("Vitamin C prevents colds", "Vitamin C does not prevent colds") until they are conjoined and a contradiction emerges. Then, since we are unwilling to assent to a contradiction, we revise some of our earlier views.

Perhaps accepting p and accepting q is prima facie evidence for accepting p & q, and hence, other things being equal, for believing p & q. But such evidence can be overridden by the agent's refusal to utter, inscribe, assent to, or act as though p & q. So a conjunction principle is not part of the "logic of belief". It is just not true that we always believe the conjunction of the individual statements that we believe.

Our practice in ascribing beliefs is flexible, and the explication of belief ascriptions should do justice to that flexibility. This is something that standard accounts are unable to do. They hope to discover a single rule or criterion in mind which determines the limits on permissible paraphrase in all correct ascriptions of belief. I doubt that there is any such rule. For the limits on paraphrase vary with variations in circumstances which such explications purposely ignore. Any single rule is apt to be too liberal to accommodate some cases or too restrictive to accommodate others.

There is a family resemblance among our ascriptions of belief, but the family is an extended one with any number of odd and embarrassing relatives. Most of the difficulties in the ascription of belief concern mistaken beliefs of one sort or another. And it is unlikely that we will be able to design or discover a single rule that anticipates and accommodates all the ways beliefs can go wrong. Instead of such a rule, I have proposed a schema that enables us to say how individual belief ascriptions function. It recognizes that such ascriptions differ in translucence, so that there is no saying in general whether in paraphrasing a belief ascription we ascribe the same belief or a different one. We have a choice as to how broadly or narrowly to construe the classification condition. And in different circumstances, different choices may be appropriate. But we cannot construe it any way we please. Any selection we make is defeasible. If, relative to our interests, a different classification condition or a greater or smaller range of paraphrases yields a better account, then the classification condition we have chosen must be replaced. And if there is no general rule for telling which accounts are better, that is because of the multiplicity and variety of roles that our ascriptions of belief play in our understanding of ourselves and of each other.

To the controversy in the philosophy of mind over the nature, character, or essence of beliefs, I have contributed nothing. The term "belief" may, for all I have said, denote a physical state, a functional state, an immaterial mental state, or some as yet unimagined alternative. But whatever the nature of such entities, there remains the question of their classification. And there is no more reason to think that mental entities determine their own classification than there is to think that physical entities do. In different circumstances it might be correct to call a physical object a chair, a Hepplewhite, an antique, or an investment. Likewise, in different circumstances it might be correct to call a single mental state a kangaroos-are-carnivores-sentence, a kangaroos-are-meat-eaters-sentence, or a marsupials-of-the-family-macropodidae-are-meat-eaters-sentence. Whether a cognitively favored sentence is identical with or is the counterpart of a mental entity, we still need to know how it is to be classified and under what circumstances it is true. It is the ways we go about answering these questions that my schema is designed to illuminate.

CHAPTER 7

Mainsprings of Metaphor

Catherine Z. Elgin with Israel Scheffler

Josef Stern dismisses extensional theories of metaphor on the ground that substitution of coextensive terms does not always preserve metaphorical truth. "It may be a truism that the metaphorical depends on the literal, but this cannot mean that the extension of a term interpreted metaphorically simply depends on its extension interpreted literally."[1] True enough. But extant extensionalists are committed to no such thesis. To say that metaphor is determined extensionally is not to say that a term's metaphorical extension is determined solely or simply by *its* literal extension. Rather, the extensional references of associated literal and metaphorical expressions intertwine in fixing a term's metaphorical reference. Among the resources available to the extensionalist are the interpretations of related literal and metaphorical expressions[2], secondary extensions[3], mention-

---

[1] Josef Stern, "Metaphor as Demonstrative," *Journal of Philosophy* 82 (1985): 683–684. Stern replied to the comments in this chapter in Josef Stern, "Metaphor without Mainsprings: A Rejoinder to Elgin and Scheffler," *Journal of Philosophy* 85 (1988): 427–438.

[2] Nelson Goodman, *Languages of Art* (Indianapolis: Hackett, 1976), 71–80.

[3] Nelson Goodman, "On Likeness of Meaning," in *Problems and Projects* (Indianapolis: Hackett, 1972), 221–230; "On Some Differences About Meaning," ibid., 231–238; *Languages of Art*, 204–205; *Ways of Worldmaking* (Indianapolis: Hackett, 1978), 104.

selection[4], exemplification[5], and complex reference.[6] There are no recipes for determining metaphorical meaning. But there are heuristics that guide our search, providing cues and clues about which aspects of the context and background might be relevant.

Stern notes that substitutivity of literally coextensive terms fails to preserve metaphorical truth. Although

(A)   Juliet is the sun.

is true,

(B)   Juliet is the largest gaseous blob in the solar system.

is false. Since the sun is the largest gaseous blob in the solar system, Stern concludes that extensionalism fails. He summarily dismisses secondary extensions, failing to appreciate the firepower they add to the extensionalist's arsenal.

A secondary extension of a term is the (primary) extension of a compound containing that term. The extension of "sun-description" is thus a secondary extension of "sun". But the extension of "sun-description" is not determined by that of "sun". Some sun-descriptions, such as those found in mythology, astrology, and ancient astronomy, are not true of the sun. Still, "sun-description" has a determinate extension—a particular class of words and phrases. And although we have no rule for its instantiation, sun-descriptions are readily recognized. Indeed this state of semantic affairs is common: we have no rule for the instantiation of "chair" either, but we recognize chairs without difficulty. To be sure, we probably cannot decide every case. We may be hard put to tell whether an odd linguistic construction is a sun-description, and whether an odd material construction is a

---

[4] Israel Scheffler, *Beyond the Letter* (London: Routledge and Kegan Paul, 1979), 31–36 and 142 n. 97; "Four Questions About Fiction," *Poetics* 11 (1982): 279–284.

[5] Goodman, *Languages of Art*, 50–58 and 85–95.

[6] Nelson Goodman, "Routes of Reference," in *Of Mind and Other Matters* (Cambridge: Harvard University Press, 1984), 61–66; for further discussion of this and the previously mentioned devices, see also Catherine Z. Elgin, *With Reference to Reference* (Indianapolis: Hackett, 1983), 43–50, 71–81, 146–154.

chair. But such difficulties do not impugn the determinacy of either extension.

Terms that agree in primary extension typically disagree in secondary extension. Although the primary extensions of "sun" and "largest gaseous blob in the solar system" are the same, their secondary extensions are not. "Apollo's flaming chariot", for example, belongs to the secondary extension of "sun", but not to that of "largest gaseous blob in the solar system". Literal meaning, Goodman suggests, is a matter of primary and secondary extension. So coextensive terms that differ in secondary extension also differ in meaning.[7] It follows that if metaphorical interpretation is a function of literal meaning, coextensive terms with different secondary extensions bear different metaphorical interpretations. Even if Romeo was an extensionalist then, his assertion of (A) did not require him to accept (B). Extensionalists are committed to basing interpretation on nothing but extensions. But we are free to invoke any, and as many, extensions as we like.

A more perplexing problem is that the single term "sun" itself bears disparate metaphorical interpretations. Juliet is characterized as the sun because she inspires passionate love; Achilles, because he is prey to awful fury. Metaphorical meaning, it seems, attaches to tokens, not types.

But how do literally coextensive replicas—that is, tokens of a single type—differ semantically? Mention-selection provides the answer. In a mention-selective application, an expression refers, not to what it denotes, but to mentions thereof. "Centaur" mention-selects centaur-descriptions and centaur-pictures; "sun" mention-selects sun-descriptions and sun-pictures. But not every sun-description need be mention-selected by a given inscription of "sun". Tokens occurring in a work of Ptolemaic astronomy, for example, mention-select "moving celestial body"; tokens in a work of Copernican astronomy mention-select "motionless celestial body". Literally coextensive replicas thus can, and often do, diverge in mention-selection. Still, mention-selection is extensional. There is a determinate class of expressions mention-

[7] Goodman, "On Likeness of Meaning," 227.

selected by each token; and two tokens are co-mention-selective just in case they mention-select exactly the members of the same class.

Replicas that bear disparate metaphorical interpretations differ in mention-selection. Inscriptions of "sun" that apply metaphorically to Juliet mention-select expressions such as "life-sustaining", and "beauteous"; those that apply to Achilles mention-select expressions such as "life-threatening" and "terrifying". The associations effected by the metaphorical application of an inscription are thus products of the literal mention-selective reference of that inscription. It is important to note that coextensive terms that differ in secondary extension also differ in mention selection. Unlike "munchkin", "wizard" mention-selects "wise man". So "wizard" applies metaphorically to people to whom "munchkin" does not.

Metaphors achieve their effects through likening. This involves yet another mode of reference—exemplification. If a symbol both refers to and instantiates a label, it exemplifies that label. And if two symbols refer to and instantiate exactly the same labels, they are coexemplificational. For example, a paint sample that both refers to and instantiates "vermilion" exemplifies "vermilion"; and separate paint samples that refer to and exemplify exactly the same labels—say, "vermilion", "flat", and "latex"—are coexemplificational. Exemplification, like the other modes of reference we have considered, is thus extensional.

Things that do not ordinarily function as symbols come to do so by serving as samples of, and thereby exemplifying, labels they instantiate. This is the key to metaphorical likening. In calling Juliet "the sun," Romeo highlights features she shares with the (literal) sun. Through his characterization, he brings her to exemplify labels such as "glorious" or "peerless". So in the simplest case, a chain of reference links the literal and metaphorical extensions of a term via their joint exemplification of a label. Thus Sol, the literal referent of "sun", and Juliet, the metaphorical referent of "sun", are linked by their joint exemplification of the label "glorious". Longer and more complex chains may also connect literal and metaphorical subjects. Juliet might exemplify a label that exemplifies a label that . . . is exemplified by the sun. And the labels exemplified may be literal or metaphorical. Any number of chains can be operative at once. A rich metaphor is inexhaustible in that additional chains of reference between its subjects may yet be forged.

Context influences the interpretation of metaphors in several ways. Normally a term is applied as part of a scheme of implicit alternatives. And a single expression might belong to a number of schemes. "Night", for example, can be opposed just to "day", or to "morning", "afternoon", and "evening". Moreover, the extension of a token of "night" varies slightly depending on which scheme is in play. Deep twilight belongs to the extension of "night" under the first scheme, to the extension of "evening" under the second. Settling the interpretation of a given token thus involves determining what kindred terms are, or might be, used in a given context. This is so whether the token functions literally or metaphorically. Rosalind is easily recognized as the referent of Romeo's "moon", for the way has been paved by his calling Juliet "the sun".

Interpretation often depends on precisely what words are used and how they are described. This is largely a contextual matter. Romeo's previous Juliet-descriptions and Rosalind-descriptions, along with his conduct vis-à-vis the two women, reveal that his love for Juliet has totally eclipsed his affection for Rosalind. So his metaphorical comparisons of the two may reasonably be expected to favor Juliet. We do not need his explicit avowal of (A) to recognize Juliet as the proper referent of "sun", Rosalind as that of "moon".

Candidacy for exemplificational reference may also be circumscribed by contextual factors. The options for a correct interpretation of (A) are limited by the fact that it is uttered by a love-sick adolescent. We can expect the labels jointly exemplified by Juliet and the sun to be superlatives appropriate for describing objects of love and desire. They are unlikely to be predicates that belittle their referents. So even though the sun is in fact a relatively insignificant star, contextual considerations exclude "insignificant" as a contender for joint exemplification. Whatever Romeo is getting at in calling Juliet the sun, he is not conceding that in the greater scheme of things she is relatively unimportant. Such contextual considerations are plainly insufficient to determine exactly which labels are jointly exemplified. But by restricting the candidate pool, they focus our search, directing our attention to a neighborhood in which a correct interpretation might be located.

Our primary objective here has been to defend extensionalism against Stern's cavalier dismissal. The account we sketch turns out to be stronger than the one Stern proposes. Unlike Stern, we invoke no intensional entities, so our theoretical basis is more economical than his. And if it accomplishes as much, the theory with the weaker basis is the more powerful. But in fact our theory explains more, and more that we particularly want to know, about metaphor. According to Stern, the most that semantics supplies for the interpretation of a metaphor is the advice: Look to context.[8] It does not explain how a metaphor likens the literal and metaphorical referents of a term. This, for students of metaphor is a (perhaps *the*) crucial question. And it surely seems to be a question about how metaphors function linguistically.

By recognizing that exemplification is a mode of reference, that words and other symbols are among the referents of our terms, and that secondary extension and mention-selection depend on actual usage, not "lexical meaning", we can explain both metaphorical reference and metaphorical likening. Unlike Stern's theory, ours makes no use of a distinction between linguistic knowledge and collateral information. This strikes us as a good thing.

[8] Stern, "Metaphor as Demonstrative," 697–698.

CHAPTER **8**

Index and Icon Revisited

Charles Sanders Peirce's legacy to contemporary semiotics is a mixed blessing: a blessing in acknowledging a variety of types of sign and modes of reference; mixed in fostering a segregation of competences and of the institutions and activities within which signs of different sorts operate. I want to unmix the blessing. My reconstrual is motivated not only by the conflict of the faculties engendered by our Peircean legacy, but also by difficulties inherent in Peirce's partition of the domain of signs. My concern is with how we should understand signs, not with how we should understand Peirce. So the legacy I describe may seem impoverished when compared with the wealth of insight to be found in Peirce's theory. Nevertheless, I believe it is that relatively impoverished legacy that influences current semiotic theories.

A sign's status as icon, index, or symbol derives from its mode of reference. Icons refer by resemblance or, as Peirce said, "mere community in some quality". Indices refer by a natural correlation or "correspondence in fact".[1] Symbols refer by convention.[2] A portrait is considered an icon then, because its reference is se-

[1] Charles S. Peirce, *Collected Papers*, ed. Charles Hartshorne and Paul Weiss (Cambridge: Harvard University Press, 1931), 1:295.
[2] Charles S. Peirce, *Collected Papers*, ed. A. W. Burks (Cambridge: Harvard University Press, 1958), 8:228.

cured by its likeness to its subject. A symptom is an index because it in fact corresponds to a disease. And most denoting terms are symbols in Peirce's sense, for their relation to their objects is a matter of arbitrary convention.

By projecting Peirce's criteria from such familiar examples, we develop a sense of what the several signs are, where they are to be found, and how they are to be understood. Since terms are paradigm cases of symbols, we readily infer that language is conventional and that linguistic competence lies in the mastery of arbitrary rules and correlations effected by a particular speech community. Portraits are obvious examples of icons, so it is easy to conclude that resemblance grounds representation. We understand a picture then when we can tell what it looks like. And because falling barometers, rising blood counts, and the like are exemplary indices, they belong in the lap of natural science. Wielding such signs requires factual knowledge; their connection to their objects is the fruit of empirical inquiry.

If we follow this line, rifts emerge along disciplinary boundaries. The natural is divorced from the conventional, and both from the representational, for different signs and competences are peculiar to each. Science requires a command of facts and laws to underwrite its signs; language and art can apparently thrive in a fact-free, lawless environment. Art is segregated from science. For the similarity of icon to object may be distinctive and unrepeatable; but indices rely on regularities to fix their referents. Pictures and words part company as well. You should be able to tell what an icon refers to just by looking; but you can not interpret a symbol without access to the relevant conventions. By their signs then ye shall know them. From a simple, elegant taxonomy of signs, we slide practically unawares into a stereotypical vision of everyday language, science, and art as mutually irrelevant if not antagonistic to one another.

The slide, of course, is not inevitable, nor is it mandated by Peirce's classification (though some elements of the picture are supported by his discussion). But it is by now so widely accepted that we are apt not to notice that the examples bear other interpretations and that the criteria support other avenues of projection. In heading blindly down the path of least resistance, we settle for a stereotype and overlook alternatives. The grip of this stereotype can be broken, I suggest, if we consider the ways signs (including those in the familiar examples) actually function.

A critical difference between symbols and other signs may be this: icons and indices would bear the same relations to their objects whether they were interpreted as doing so or not. The similarity of portrait to subject, like the correlation of symptom with disease, obtains independently of anyone's taking the one to be a sign of the other, indeed, independently of anyone's recognizing a connection between the two. But without a linguistic community whose members attached a particular interpretation to the word "dog", there would be nothing to correlate "dog" with dogs. A convention thus *makes* the connection between symbol and referent; it does not merely deploy a connection that already obtains. Unlike icons and indices, symbols are related to their referents only because they are so interpreted.

The difficulty is that resemblances and natural correlations are ubiquitous. Every two entities bear some likeness to each other, and some correspondence in fact. Yet we do not consider every object a sign, much less an icon or an index of every other. Umberto Eco is hardly an icon of La Scala, despite their being alike products of Italy. Nor is a full moon an index of an avalanche, though in fact the moon was full when the avalanche occurred.

Something is an icon or an index only if it functions as such. His mug shot is an icon of the criminal, and a fever an index of an illness because they are taken to signify their objects. But being taken to signify requires an interpretant. So icons and indices, like conventional signs, are symbols. If a distinction is to be drawn, it must be within the class of symbols, not between signs that are symbols and signs that are not.

Our predicament is this: since sign S refers to object o, S is a symbol for o. Moreover, like every other object, S resembles o. And like every pair of objects, S and o bear some correspondence in fact. Icon, index, and symbol threaten to collapse into an undifferentiated heap. For any sign that satisfies the conditions on one apparently satisfies the conditions on all.

Isn't there this difference? What Peirce calls symbols are supposed to be *purely* conventional. Anything could, in principle, symbolize anything else, provided the requisite conventions were in place. But icons and indices require more: a nonconventional hook to their referents. This does not yet help; for the requisite nonconventional relations always obtain. If every two objects are alike in some respect, then with the help of the right conventions, any object could be an icon of any other. And if every two objects are connected in fact, then backed by appropriate conventions, any object could be an index of any other.

Still, we are in the right neighborhood. If we retreat from counterfactuals, we may see our way more clearly. The issue is not whether a given symbol *S* could be an icon or index of *o*, but whether it is one. *S*'s being a symbol of a particular kind, as much as its being a symbol at all, is determined by the way it functions. Whether *S* is an icon of *o* depends not just on whether there is a resemblance between *S* and *o*, but on whether *S* refers via that resemblance. And whether *S* is an index of *o* depends not just on whether there is a natural connection between the two, but on whether *S* refers to *o* via that connection. If not, the resemblance and the natural connection, though real, are semiotically inert. Unlike purely conventional symbols then, icons and indices use nonconventional links to their objects for referential purposes.

To make anything of this, we need to know what referring via a feature involves. Certain devices introduced by Nelson Goodman provide the key. Exemplification, Goodman contends, is the mode of reference that links a sample to what it is a sample of. A fabric swatch, being a sample of pattern, color, texture, and weave, refers to its pattern, color, texture, and weave. The swatch is not, ordinarily, a sample of its size and shape; for although they are as much its properties as its pattern and weave, the swatch makes no reference to them. Exemplification requires reference as well as instantiation.[3]

By means of referential chains, symbols may refer indirectly.[4] A tiger, for example, might refer to and be a symbol of a team by

---

[3] Nelson Goodman, *Languages of Art* (Indianapolis: Hackett, 1968), 53.
[4] Nelson Goodman, *Of Mind and Other Matters* (Cambridge: Harvard University Press, 1984), 62–63; Catherine Z. Elgin, *With Reference to Reference* (Indianapolis: Hackett, 1983), 142–154.

exemplifying, say, ferocity, a characteristic also possessed by the team. Not all features shared by the tiger and the team, however, constitute links in a referential chain. Even if both are expensive to maintain, *expensive to maintain* is unlikely to forge a referential chain linking the two.

What I want to suggest is that to refer to an object *via* a feature is to refer to it by means of a referential chain that has the exemplification of that feature as an intermediate link. An index then refers via a regularity only if it refers to its object by exemplifying a regularity that the object also instantiates. Only in contexts where it exemplifies nervousness then, does stammering serve as an indexical sign that a speaker is nervous. The connection in question need not, of course, involve a recondite law of psychology. If established in fact, something as commonsensical as "Nervous speakers often stammer" fills the bill. Similarly, an iconic sign refers via a likeness only if the exemplification of that likeness serves as an intermediate link in a referential chain connecting symbol and referent. If her portrait is an icon of the queen, it refers to the queen by exemplifying features it shares with her.

Is this enough? It may do for the explication of indices; but as it stands, it seems to count too many symbols as icons. We do not mind classifying her portrait as an icon of the queen if it exemplifies features like the shape of her brow and the tilt of her chin. But we would be reluctant to treat a political cartoon depicting homeless people sleeping on grates in front of the White House as an icon of Ronald Reagan, even though it refers to him by the exemplification of a feature like *callous indifference to need* which he evidently shares. Minimally, it seems, an icon of Reagan should depict or otherwise represent Reagan.

To complete our explication of icons then, we need to invoke a third Goodmanian device—secondary extension.[5] In addition to its primary extension, Goodman maintains, a denoting symbol has a number of secondary extensions. These are extensions of compounds containing the symbol. The primary extension of the word "house" consists of houses. One secondary extension consists of house owners, another of housewares, the terms "house

---

[5] Nelson Goodman, "On Likeness of Meaning," in *Problems and Projects* (Indianapolis: Hackett, 1972), 227.

owner" and "houseware" being compounds of "house". Some of a symbol's secondary extensions depend on its primary extension; others do not. In particular, the extensions of compounds like "house picture", "house description", and more generally, "house representation" are independent of the extension of "house". A fictive house picture denotes no house; and a photograph of a square foot of blistered paint, though, regrettably, a picture denoting my house, is no house picture. Whether a picture qualifies as a house picture is determined by our classificatory practice. And that practice is not determined, though it may be influenced, by the denotation of the pictures it pertains to.

I suggest an icon of Reagan must belong to a secondary extension of the name "Reagan". A work is a Reagan icon then only if it is a Reagan-representation that refers to Reagan by exemplifying features it shares with him. A picture depicting Reagan as indifferent to need could be an icon of Reagan; a picture depicting only the victims of his indifference could not.

How can such a suggestion be evaluated? We cannot very well demand agreement with an antecedently accepted standard of iconicity. We have none. Still, the term "icon" has a use in our language. We readily classify some signs as icons and refuse to so classify others. Yet others fall into a third category—the "don't knows". An adequate explication should accommodate this classification. Its verdicts should, for the most part, agree with those we are inclined to give on cases we consider clear, and should settle cases whose status is antecedently in doubt.

A fairly modest requirement, this. Unfortunately, my proposal does not satisfy it. For we readily classify as icons both fictive symbols and symbols whose referents are epistemically inaccessible. In neither case can exemplification of features shared by their objects account for their iconicity.

Santa-Claus-pictures cannot exemplify features their object shares, for they have no object. And we have no way to tell whether the features a Saint-Jerome-picture exemplifies are shared by their object. Saint Jerome is long gone; his features are unavailable for comparison. Yet we have no qualms about calling pictures of both sorts icons.

It will not do to retreat to Peirce's original proposal. For it is vulnerable to the same difficulty. Santa-Claus-pictures can no more resemble their object than they can exemplify features of it. And Saint-Jerome-pictures can no more be known to resemble their object than they can be known to exemplify features he shares.

We are not entirely without resources, though. We saw that a work qualifies as a Santa-Claus-picture on account of its secondary, not its primary, extension. It qualifies, that is, because it is appropriately related, not to Santa Claus, but to other Santa-Claus-representations. Similarly, I suggest, it qualifies as a Santa-Claus-icon because a suitably configured referential chain links it to that secondary extension. More precisely, a Santa-Claus-icon is a member of the class of Santa-Claus-representations that refers to the membership of that class by exemplifying features some other members also exemplify.[6] Obvious candidates for joint exemplification include features like being portrayed as jolly, chubby, white-haired, and gift bearing. But these particular intermediate links are not necessary and not always sufficient to complete the referential chain.

Icons of inaccessible objects refer by means of a similarly structured chain. We cannot tell whether a Saint-Jerome-representation exemplifies features shared by the historical Saint Jerome, but we can tell whether it exemplifies features exemplified by other such representations. So if a Saint-Jerome-icon's referential chain passes through its secondary extension, each link is in principle accessible. A Saint-Jerome-icon is, I suggest, a Saint-Jerome-representation that refers to Saint Jerome by exemplifying features likewise exemplified by other Saint-Jerome-representations that denote Saint Jerome.

Although the works in question must denote Saint Jerome, they need display no affinity to him. The connection of a denoting symbol to its object may be utterly arbitrary as, for example, the connection of a name to its referent typically is. So a picture's denoting Saint Jerome is semantically independent of any likeness it may bear to him. If, as seems likely, Saint Jerome was not in fact befriended by a lion, pictures portraying him as such are

---

[6] The step from some members to the class involves synecdoche.

inaccurate. Still, they may be Saint-Jerome-icons. For that requires that they exemplify features of other Saint-Jerome-representations, not that they exemplify features of the historical figure, Saint Jerome. Since the referential link to Saint Jerome himself is a matter of convention, a work's infidelity to fact does not impugn its status as icon.

Evidently no general rules specify what sorts of features an icon must exemplify. This is somewhat surprising. We might expect icons of a single subject to agree about the subject's inherent perceptible features. In that case, though the circumstances in which they place him might vary, pictorial icons of Saint Jerome would look like portrayals of the same man. But if we examine the works we are inclined to call icons, we see otherwise. We find tall Saint Jeromes and short ones, well-fed and ascetic ones, gloomy Saint Jeromes and sanguine ones. Icons may effect reference by exemplifying not their subjects' inherent features but, as it were, the trappings of his office. Saint-Jerome-icons are apt to agree in exemplifying that they portray, for example, a scholar, a hermit, a Church father, and/or a companion to a loyal lion. The relevant "trappings of office" are, of course, matters of convention. They vary from case to case, and change as tradition evolves. They may, moreover, include fictional and metaphorical, as well as factual and literal features, and may involve matters of style as well as substance. Plainly this is not the sort of thing an uninformed viewer could recognize in a work just by looking at it.

The icons I have cited as examples have all been pictures, these being, perhaps, the most familiar and uncontroversial cases. But neither Peirce's discussion nor common parlance provides justification for restricting the iconic to the pictorial. We would hardly deny that a statue of Saint Jerome is an icon merely because it has three dimensions rather than two. Nor are models or diagrams obviously excluded. Indeed, it's not even clear that we ought to restrict the iconic to the visual. Mimicry might, for example, refer iconically. If so, language, gesture, and music admit of icons as well.

To allow for this possibility, I have taken the relevant secondary extension to consist of representations of a subject, rather than merely pictures of it. And representations may belong to any symbolic medium. My choice is not question begging, though. For if my explication is correct, whether verbal, musical, or gestural signs function iconically remains an open question—one that is to be decided by seeing whether they refer via the proper sort of referential chain. The answer thus depends on the uses we make of such signs, on their functions in a symbol system. It cannot be settled a priori.

Interpretation is required to fix each link in a referential chain. Whether a given work is a Saint-Jerome-representation, a Santa-Claus-representation, or a Queen-Elizabeth-representation depends on our practice of classifying works, a matter of convention that varies significantly from one case to the next. Only by (implicitly or explicitly) invoking the appropriate conventions can the question be decided. Whether the work exemplifies features that its subject shares or features that others of its kind also exemplify likewise requires interpretation. Having features in common is not enough. Exemplification requires reference as well. And although we can sometimes tell just by looking whether a picture and its subject or several pictures of the same subject share certain features, we cannot so easily tell whether those features are exemplified. For that we need to know what and how the presence of those features contributes to the symbolic functioning of the works they belong to. This depends on how the conventions applicable to the works are deployed and how the options those conventions leave open are exercised.

The existence of a referential chain is not insured by the existence of its several links. The links must be connected. Even if, for example, a given Saint-Jerome-picture exemplifies a feature that other such pictures also exemplify, it may fail to refer to those pictures, or fail to refer to them via the exemplification of that particular feature. A chain exists only if the picture in question uses the joint exemplification of that feature as a vehicle for referring to the other works. If not, their exemplifying the same feature might be a coincidence or a stylistic or thematic affinity, but it would not be a link in a referential chain.

Interpretation is always required to decide whether and how a symbol functions as an icon. Since the routes of iconic reference are intricate and the identity and connectedness of the intermediate steps often controversial, interpreting such signs is a delicate and sometimes difficult business.

No medium or discipline has exclusive dominion over any sort of sign. Pictures and words may function as diagnostic cues—as indices of medical or psychological conditions. Even though the spontaneous utterance of obscenities is typically a conventional response to an unwelcome situation, it is sometimes an index of Tourette's syndrome—a neurological affliction. Despite the fact that the signs are linguistic, why and how this particular connection in fact obtains is a problem for science. Words and natural signs may function as icons as well. One speaker may refer iconically to another by mimicking his style of speech. A transient ischemic attack may refer iconically to the stroke it portends by exemplifying stroke-like characteristics. And pictorial and natural signs may be conventional symbols. A halo is by convention a pictorial sign of holiness; a groundhog's sighting his shadow, a natural sign that winter is not yet over. The difference between pictures, signals and descriptions thus is not, and is not derivable from the difference between iconic, indexical, and purely conventional signs.

Peirce's discussion suggests that icons and indices immediately and directly signify their objects, that they are simpler and more easily grasped than conventional symbols. We see now that this is not the case. Nor is it obvious that the notions of icon and index retain any theoretical utility once the suggestions of simplicity and directness are abandoned. Referential chains admit of a variety of configurations. The continued utility of the notions of icon and index turns on whether the chains whose configurations I have identified have any special theoretical standing. There is currently no reason to think that they do.

Whether or not these particular notions endure, however, the semiotic devices that enter into their explication are important. They enable us to recognize that reference may be highly complex, and that a single sign may at once stand in several referential relationships to several different referents.

CHAPTER 9

Sign, Symbol, and System

The Monday-morning quarterback reports: "The fullback headed for a hole that rapidly closed, bounced a couple of yards to the left, scrambled through a snarl in the defensive line, and flew down the field for a touchdown." She illustrates her story by manipulating the breakfast crockery. Her coffee mug represents the fullback; the sugar bowl, some cereal dishes, and a juice glass stand for the defensive line.

Such reports are so familiar that we easily overlook their symbolic complexity. The report contains two representations of the play: the verbal description and the illustration effected by manipulating the crockery. The description belongs to standard English—a sophisticated, complicated, widely known, and well established symbol system. The illustration is reasonably crude and ad hoc. The two representations are independent of each other. Once interpretations have been assigned to their referring elements, neither requires appeal to the other to represent the play. The Monday-morning quarterback can describe the play without illustrating it. And she can slide her (interpreted) dishware around the table, illustrating the play without saying a word. By comparing the two representations, we may discover some interesting features of symbols and the systems they belong to.

Since the term "symbol" is ambiguous, we should settle its interpretation first. According to one usage, symbols indicate through indirection. Thus a psychiatrist might take her deployment of the *coffee mug* to be highly symbolic of the Monday-morning quarterback's neurotic fixation. A cultural anthropologist might maintain that football is a form of symbolic warfare. This is not the interpretation I am concerned with. I use the term "symbol" to denote anything that refers—directly or indirectly. The coffee mug and the expression "the fullback" then both symbolize the fullback. And the term "football game" directly and literally symbolizes football games, whether or not football games indirectly and figuratively symbolize warfare. Plainly there are differences between the direct and the indirect, the obvious and the arcane, the pedestrian and the esoteric. But these are differences among symbols, as I use the term, not between things that are symbols and things that are not.

The coffee mug and the expression "the fullback" then refer to the same individual. So do a variety of other symbols. We call him by name, by the number on his jersey, and by a host of laudatory and opprobrious epithets. And in their reconstructions and revisions of crucial plays, Monday-morning quarterbacks are prepared to take practically anything to designate the protagonists. None of this is particularly surprising, but it is worth attending to.

Picture theories of language, and iconic theories of pictorial representation maintain that symbolizing depends on similarity of sign and thing signified. Because A is (appropriately) similar to B, A refers to B. Just how A and B are required to resemble each other is not obvious. But the similarity in question must be a natural, rather than a conventional, relation.[1]

Certain difficulties are well known. Resemblance is symmetrical; reference is not. The fullback resembles the coffee mug as much and in the same respects as the mug resembles the man. But the mug represents the fullback, the fullback does not represent

[1] See Ludwig Wittgenstein, *Tractatus Logico-Philosophicus* (London: Routledge and Kegan Paul, 1961); Wilfrid Sellars, *Science and Metaphysics* (London: Routledge and Kegan Paul, 1968), chapter 5; and Jay F. Rosenberg, *Linguistic Representation* (Dordrecht: Reidel, 1975) for views of this kind.

the mug. The mug in question doubtless resembles other mugs more closely than it resembles any football player. Still, in this context it does not refer to other mugs.[2] The first of these difficulties might be overcome by defining a suitable asymmetric relation on ordered pairs of relevantly similar objects. To overcome the second requires identifying the respect in which sign and thing signified must be alike.

A given coffee mug and a large, fast athlete have much in common. Many shared features are manifestly irrelevant. Although both mug and man are subject to gravity, this characteristic is too widely shared to serve. The relevant resemblance must be selective. Since the mug represents the man intermittently, enduring similarities are ruled out as well. The relevant resemblance obtains only when the mug represents the athlete. Perhaps the path of the mug across the table reflects (closely enough) the path of the athlete down the field. If this hypothesis is to be taken seriously, the notion of "reflecting (closely enough)" should be investigated in detail. But such investigation is unnecessary because the hypothesis is easily shown to be implausible.

The mug traverses a particular path during the illustration of the play. Then the Monday-morning quarterback retrieves it, finishes off her coffee, and pushes it across the table for a refill. On both occasions the mug traverses the same path. But only on the first does it represent the fullback. The resemblance between the path of the mug and the path of the runner cannot secure reference. For the mug's second trip across the table resembles the path of the fullback as closely as the first. And on its second trip, the mug does not refer.

Nor is it plausible to maintain that I have simply picked the wrong resemblance. No enduring affinity will do, for the mug does not permanently signify the athlete. Yet we cannot choose some indexical feature that the mug has during the illustration and at no other time in its history. For surely the mug might again be used to symbolize the fullback. And it is reasonable to expect the same relation to account for its doing so.

The same holds for the linguistic expression "the fullback".

[2] Nelson Goodman, *Languages of Art* (Indianapolis: Hackett, 1976), 3–5.

Any natural likeness obtains whether the expression is interpreted or not. But it does not symbolize the athlete unless assigned him as its referent. Under a different interpretation the same configuration of letters does not refer to our man. The word "case", for example, sometimes denotes cartons, sometimes legal disputes; the letters *c, h, a, t* spell an English word that refers to conversations, a French word that refers to cats. No natural affinity connects a term more intimately to one referent than to another.

Instead of seeking a resemblance between sign and thing signified, perhaps we should seek an affinity among codesignative signs. This will not tell us how reference is secured, but it may reveal what makes for coextensiveness. Then we can conclude that whatever is required for reference, signs are coextensive just in case they bear the proper sort of resemblance to one another.

Like resemblance, coextension is reflexive and symmetrical. Unlike resemblance, it is transitive. If A and B are coextensive, and B and C are coextensive, so are A and C. Since the words "gato" and "Katze" are coextensive, and "Katze" and "cat" are coextensive, so are "gato" and "cat". But if A resembles B and B resembles C, it does not follow that A resembles C. Snow tires and bagels are alike shape. Bagels and croissants are alike in nutritional value. But snow tires and croissants are alike in neither shape nor nutritional value. It is not enough that each symbol that refers to a given object somehow resemble all others that do so. Rather, all such symbols must bear the same resemblance to one another.

Our prospects of finding such a resemblance are bleak, particularly if we restrict ourselves to natural likenesses, which are supposed to obtain apart from any conventions we use, any devices we employ, any interpretations we impose, any terminology we introduce. The coffee mug and the expression "the fullback" are bound to have something in common. But, as we saw earlier, this is not enough. For the individual in question is also designated by expressions like "number 80", "Freeman", and "Slippery Sylvester", as well as variety of descriptions that ought not be uttered in polite society. And Monday-morning quarterbacks might represent him diagrammatically by means of objects as diverse as furniture, flower pots, and fruit. It is unlikely that we will find a natural affinity among all and only those things that symbolize him, particularly since many of the symbols—coffee mugs, for example, and expressions like "that incompetent excuse for an athlete"—are apt to designate him temporarily, and abruptly cease to do so.

Suppose we abandon our quest for a natural likeness and consider whether some conventional resemblance accounts for coextensiveness. In that case, we give up the attempt to explain the symbolic in terms of the nonsymbolic. But we may still be able to explain sameness of reference in terms of other, better understood, semantic features.

Might symbols be coreferential because they perform the same function in their respective symbol systems? If so, the coffee mug and the expression "the fullback" are codesignative because their symbolic roles are the same. And when the mug ceases to play that role, it ceases to bear the requisite functional resemblance to the term. As a result, it no longer designates what the phrase "the fullback" does.[3]

The difficulty comes in explicating the notion of sameness of symbolic function. How are we to tell whether signs that belong to distinct systems play the same role? There are seemingly important differences between the signs we are considering—differences that stem from the fact that they belong to quite different systems.

The mug's role as a symbol is transient and ad hoc. It need have no history of denoting the fullback, nor is there implicit in its present application any expectation that it will continue to do so. Should the Monday-morning quarterback, on some future occasion, again take the mug to denote that player, she will be obliged to re-establish the referential connection. Certainly she can do so. And if the same dishes are regularly assigned the same referents,

---

[3] This proposal is reminiscent of the semantic theory of Wilfrid Sellars. See *Science and Metaphysics*, chapter 3. But Sellars restricts his account to linguistic systems, whereas I want to consider symbol systems in general. And he hopes that sameness of symbolic function will serve as the basis for a relation of synonymy, whereas I am concerned to identify the basis of the broader notion of sameness of reference.

a more or less autonomous symbol system might develop. Its referential relations will become relatively independent of any particular occasion of use. But such an autonomous system is not required for the original illustration to be effective.

The expression "the fullback", of course, belongs to an enduring symbol system—the English language. That the term refers to Freeman, or to the player in the backfield with certain specifiable duties might be explicitly stated to inform a novice. But because the term has a secure, stable place in the language, its reference need not be re-established each time it is used. The expectation that it will continue to refer to what it did in the past is implicit in the nondeviant use of the term. The referential connection between word and object is then reasonably independent of any single utterance or inscription.

The linguistic expression is woven into a rich and complex syntactic and semantic network. The mug is not. English has the resources to refer to the players on the bench, and to evaluate the performances of different players. It has the means to compare the position of fullback in football to its counterparts in other sports, and to other positions on the football team. And it has the capacity to apply the term "fullback" metaphorically to individuals who are not engaged in athletic endeavors. The mug's symbol system lacks these resources. It might be enriched by introducing symbols and relations among symbols to perform such functions. But so far, it has none. The word then can play a number of symbolic roles that the mug cannot—simply because the mug belongs to a vastly more impoverished system.

Indeed, each of the symbols we have used to denote the player has symbolic functions the others lack. The mug symbolizes the Monday-morning quarterback's neurotic fixation. (Just how is a question perhaps best left to her therapist.) The term "fullback" makes reference to the player's location on the field before the ball is snapped. If the player is African American, his name—"Freeman"—might allude to the emancipation of the slaves. The number on his jersey recalls and invites comparison with previous team members who wore the same number. It thereby locates Freeman in an ongoing athletic tradition. Sameness of symbolic function is manifestly too much to ask.

What if we restrict our attention to literal descriptive functions? Then, perhaps, we can make some headway by replacing the requirement of sameness of function with something like correspondence of literal function. The mug and the term "fullback" are then coreferential, because their literal descriptive functions correspond. The difficulty is that we cannot give a strict and definite criterion of correspondence. And symbols in different systems may correspond more or less closely. In a system that differentiates only between backfield and line, the closest counterpart to the mug would be a symbol that refers to a backfield player. Is that close enough? There seems to be no clear answer. Without a definite criterion of correspondence, the relation of sameness of reference is indeterminate. We have no way to tell whether symbols with different functions correspond closely enough to count as coreferential.

With sufficient ingenuity we can apparently take just about anything to refer to anything. And the symbol systems we devise are apt to differ greatly in structure and representational resources. Is there anything common to coextensive symbols that justifies our calling their roles in their respective language games the same? Perhaps only the fact that they refer to the same objects. Evidently we cannot explain coextensiveness by appeal to some antecedently or independently specifiable kinship, natural or conventional. For there appears to be no way to identify the symbols as relevantly similar except by noting that they refer to the same things. Maybe we should start with that.

I don't suppose anyone has proposed a causal theory of reference to explain the relation between football players and symbols that denote them. This is just as well. For the ingenuity of Monday-morning quarterbacks raises problems for such theories. Briefly, causal theories maintain that a term's reference is fixed by stipulation and preserved from one use to the next by the intention of each speaker in a chain of communication to refer to what was referred to by earlier members of the chain, all the way back to the original stipulation.[4] Typically the theory is held to apply

---

[4] Saul Kripke, "Naming and Necessity," in *Semantics of Natural Language*, ed. Donald Davidson and Gilbert Harman (Dordrecht: Reidel, 1972), 298–299.

only to proper names and to natural kind terms—to terms like "Aristotle" and "aardvark".

Difficulties emerge. There is the problem of characterizing the causal chain that links its current utterance to the introduction of a term into the language. Aristotle, of course, was not named "Aristotle"; the name he went by had a different pronunciation and a different spelling. So the claim that our use continues the chain that began with his being baptized "Aristotle" needs refinement. Then there is the worry that chains that originate in a single stipulation may later diverge. In that case a term has two different reference classes despite its link to a single introducing event. The term "fish", for example, is sometimes but not always restricted to animals with gills. Perhaps the term was introduced in the context of the false belief that all denizens of the deep are physiologically alike. Ambiguity occurs because correction of this error allows for alternative continuations of the causal chain: one disregards habitat; the other, physiology. Each continues the chain, but the two uses of the word "fish" are not coextensive. Nor do we always succeed in referring to what our predecessors did, even when we intend to do so. A native name for a portion of the African mainland was mistakenly applied to the island we now call "Madagascar".[5] If the intention to refer to what our predecessors did determined reference, the term would still refer to the mainland. It does not.

The Monday-morning quarterback's demonstration raises yet another difficulty. For the referential relation between football players and crockery is, and is meant to be, temporary. Suppose that the following evening someone objects that the location of the salt shaker designates an ineligible receiver downfield. Even if the speaker uses the salt shaker with the intention of referring to what the Monday-morning quarterback did, he has made an error. For once the demonstration ended, the salt shaker ceased to refer. A new tacit or explicit stipulation is needed if it is to function referentially again.

Because the interpretation of the crockery as a symbol system is so obviously circumscribed by contextual factors, it affords a

[5] Gareth Evans, "The Causal Theory of Names," *Proceedings of the Aristotelian Society Supplement* 47 (1973): 195–196.

good example of the problem of overextending chains of communication. And the problem remains if we take the various dishes to be proper names of football players rather than descriptions of them in terms of their positions. It just is not true that the symbols continue to refer so long as the appropriate causal chain remains intact.

It is not only in ad hoc demonstrations that chains are overextended. A well-intended but ill-informed speaker might use the name "Myra Mulligan" to refer to the person to whom it was intended to refer by the speaker from whom he learned it, and so on, all the way back to the event at which the name was introduced into the language. Although the causal chain is intact, the Mulligans' marriage is not. The person to whom the speaker intends to refer now goes by the name of Myra Morgan. And as a result of Waldo's predilection for women named Myra, the name "Myra Mulligan" now refers to someone else entirely. To insist that because the causal chain is unbroken, the name "Myra Mulligan" continues to denote the woman who now answers to "Myra Morgan" is to invite confusion, not to say hostility. And to say that she would have been Myra Mulligan even if she had never become involved with Waldo is to say something manifestly false.

The conventions by which a woman takes, retains and/or gives up her husband's surname are complicated. And we have a certain latitude of choice regarding what name to use. But it is just wrong to suppose that once she has taken his surname, that name continues to refer to her regardless of changes in her circumstances, so long as the appropriate chain of communication remains intact. A theory of proper names that fails to do justice to the names of a sizable portion of the population seems seriously inadequate. Referring is a more complicated business than the causal theory recognizes.

Is what a symbol refers to simply a matter of convention? After all, the mug denotes the player only in the context of a particular language game and a convention was required to delimit that game and the role of each piece in it. That the mug would function differently if different conventions were in force seems obvious. And that we might easily have employed different conventions seems equally obvious. Even so, we would not be warranted in concluding that the relation of sign to thing signified is entirely conventional.

How we symbolize depends in large measure on our interests, our insights, and our ingenuity. But it does not follow that with sufficient creativity and effort we can design a system of conventions that enables us to represent things any way we please. A symbol system must be internally consistent. It must not impose incompatible labels on a single object. For the mug to represent a player who both is and is not a ball carrier is impermissible. But a system does not become consistent by our declaring it so. Nor are the only constraints logical ones. Not all goals are jointly satisfiable—even when their joint satisfaction entails no contradiction. Where they are not, tradeoffs must be made, or objectives abandoned. For the convenience of easy illustration, the Monday-morning quarterback sacrifices sensitivity. She can sketch the play, but nothing so clumsy as pushing around the crockery can capture the delicate combination of speed and grace, power and agility that characterized Freeman's run. She can opt for convenience or sensitivity, but she is unlikely to devise a system that exhibits both.

The theories we have been considering presuppose a sharp distinction between scheme and content. The realist maintains that the world (the content of our system) comes neatly divided at the joints. A scheme provides only a vocabulary for representing what is already the case. A naturalistic relation—resemblance, causality, or whatever—links scheme and content. The functionalist holds that conceptual or symbolic roles are settled independently; our labels serve only to mark them out. The conventionalist claims that by imposing different systems of conventions on a neutral, undifferentiated content, we can describe or depict the world however we like. We find in the world only the order we impose on it. And we can impose any order we please.

But the scheme/content distinction has come into disrepute, and rightly so. The orders we find are neither entirely of our own making nor entirely forced upon us. There is no saying what aspects of our symbols are matters of conventional stipulation, and what are matters of hard fact. For there are few purely conventional stipulations, and no hard facts.[6]

To be sure, a multiplicity of representations may share a subject matter. And we can learn something about symbol systems and about the subject matter by comparing such representations. But we cannot hope to factor representations into conventional and factual components, or to say independently of their various representations what the facts are.

The mug apparently has a symbolic function only when it is used as a symbol. When it is used solely as a vessel for coffee, as a paperweight, or as a flower pot, it is referentially inert. Accordingly, rather than ask "What is a symbol?" we do better to paraphrase Goodman[7] and ask "When is a symbol?" The distinction between things that are symbols and things that are not is not a permanent ontological one, fixed in the cosmic order, but varies with circumstance and purpose. The mug's status as a symbol comes and goes. It is a symbol only when it refers. It symbolizes what it refers to. And often it does not symbolize at all.

Maybe we are going too fast. The mug is a symbol only when it functions as such. But the expression "the fullback" seems to retain its semantic status from one use to the next. At least we do not need to assign it an interpretation each time we use it. Is the term, unlike the mug, permanently a symbol? Perhaps.

Items are identified as words, pictures, numerals, and the like on the basis of their symbolic properties. That is, to classify something as a word, or picture, or numeral is ipso facto to classify it as a symbol. Accordingly, it is not surprising that words, pictures, and numerals are permanently symbols. But the status of a particular configuration of ink on paper is variable. Sometimes it may be counted as a symbol, other times just as a squiggle. So, its status as word or picture or numeral varies as well. The expression "the fullback" looks to be permanently a symbol and the mug only occasionally so, not because of a difference in the ways these two items symbolize, but because of a difference in the ways they are themselves symbolized. Ordinarily the mug does not

---

[6] Nelson Goodman and Catherine Z. Elgin, *Reconceptions* (Indianapolis: Hackett, 1988), 93–100.

[7] Nelson Goodman, *Ways of Worldmaking* (Indianapolis: Hackett, 1978), 66–67.

belong to the extension of the term "symbol", whereas the word "fullback" normally does.

A symbol system determines not only the identity and classification of its component symbols, but also the identity and classification of their objects. Identification, as Goodman says, "rests upon organization into entities and kinds".[8] And that organization is effected by a constellation of constructions and devices including, in Quine's words, "plural endings, pronouns, numerals, the 'is' of identity, and its adaptations 'same' and 'other'".[9] The perennial metaphysical questions of the one and the many, and of the identity and diversity of individuals and kinds, then depend for their solutions on features of symbol systems. For these determine what counts as being the same thing, what counts as being another thing of the same kind, what counts as being something different, and what counts as being a different kind of thing.

To see this, we might take one more look at the Monday-morning quarterback's account. I have been speaking as though both the mug and the expression "the fullback" denote a specific human being. He is identified by his place on the team, but the symbols denote him on or off the field. But another interpretation is equally natural and equally consonant with her accounts—one in which the symbols represent not a particular human being, but rather a position occupied by different human beings at different times. These interpretations diverge. According to a system in which the term denotes a position, there is one fullback on a team. According to one in which it denotes a person, there are several. Under the former, there are a number of players who alternate in playing fullback. Under the latter, there are a number of fullbacks on the team, only one of whom is on the field at a time. Where the term denotes the position, a report of the game might read "The fullback was responsible for a total gain of 84 yards". Where it denotes the various players, the report reads "The fullbacks were responsible for a total gain of 84 yards". And so on.

What about the mug? We haven't singled out counterparts of

---

[8] Ibid., 8.

[9] W. V. Quine, "Ontological Relativity," in *Ontological Relativity and Other Essays* (New York: Columbia University Press, 1969), 32.

the "is" of identity, "same" and "other". But we can see whether the mug denotes the position or the player by seeing how a substitution is handled. If the mug is replaced when the man is, it denotes the man.

A symbol's ontological commitment thus depends on the symbol system it belongs to. And the symbol itself might easily belong to rival systems. Whether the Monday-morning quarterback refers to Freeman, the enduring human being, or to the time slices of Freeman and of others that are spent playing fullback, is not to be decided by examining the representation we began with, but by seeing how the relevant symbols function elsewhere.

But suppose there just is no more to the symbol system than is contained in the Monday-morning quarterback's illustration. She made up her rudimentary system to illustrate a single play; that being done, the system dies. Nothing in her demonstration decides between rival interpretations of the mug. And nothing in her intentions, or dispositions to use the mug symbolically need do so either. Since she constructed the system to illustrate a single play, her illustration might easily exhaust her intentions and dispositions regarding her symbols. And under either construal, the mug functions correctly in the context of interest. There is then no fact of the matter regarding which interpretation is correct. Since each is consonant with all the symbol's actual uses (and with the symbol user's intentions and dispositions), there is no basis in fact for preferring one to the other. Any decision must be regarded as a stipulation, not a discovery.

There can be no fact of the matter until we devise a symbol system capable of distinguishing between the rival interpretations and incorporate the mug into that system. This seems clear enough. But its implications are striking. What facts there are is a function of the symbol systems we develop. That is, we participate in the creation of the facts by creating symbol systems with the capacity to represent those facts. In Goodman's terminology, we create worlds by creating, refining, and manipulating symbols. We *create* the distinction between position and player by creating a system with the resources to draw that distinction.

We do not, of course, create facts ex nihilo. Our raw materials are the symbol systems at hand.[10] We might create a fact of the matter regarding the mug's interpretation by expanding and refining the symbol system it belongs to, or by identifying the mug with a symbol in a richer system where the relevant distinction is already drawn.

Of course, we are not at liberty to modify our old systems, or to create new ones, in any way we please. We can create no correct, literal, factual representation of the play under discussion according to which the fullback was stopped at the line of scrimmage. For, however we represent it, he made a touchdown. So any representation to the contrary is fictive, figurative, or false. And the ways fictive, figurative, and false representations symbolize is another matter—one we cannot enter into here.[11]

---

[10] Goodman, *Ways of Worldmaking*, 97.

[11] For a discussion of these matters, see Catherine Z. Elgin, *With Reference to Reference* (Indianapolis: Hackett, 1983).

CHAPTER 10

Making Up People and Things

Richard Nixon resigned. There is no mystery about why. He had lied about his role in the Watergate coverup, actively engaged in a conspiracy to subvert a congressional investigation and defied a Supreme Court ruling that he was legally obliged to obey. And he got caught.

This explanation could easily be augmented by providing more details about the actions and motives of the protagonists, the office of the presidency, the roles of Congress, the Supreme Court, and the press, and the political climate in which the events took place. Such an explanation would be shot through with intentional and cultural concepts. It would mention actions and attitudes, practices and institutions. Let us call intentional and cultural notions *anthropic* notions and accounts that employ such notions, anthropic accounts.[1] An extended explanation of the Watergate conspiracy would be anthropic. It would not restrict itself to delineating changing distributions of matter in motion.

Reductionists maintain that resort to anthropic concepts is merely an expedient. A physicalist explanation of Nixon's resig-

---

[1] This term is misleading, if nonhuman animals have intentional states. But the notions are crucial to our understanding of people. So I use it in what follows as a term of art.

nation would be hopelessly unwieldy. Economy of prose favors talking in terms of conspiracies, Congressional mandates, and political liabilities instead. If reductionism is correct, such talk is just a shorthand. It can in principle be cashed out in the vocabulary of physics with no loss of truth value or explanatory adequacy.[2] But reduction of the anthropic to the physical has failed abysmally. Political corruption evidently does not constitute a natural kind whose boundaries can be delineated without recourse to moral and political concepts. Nor is eliminativism an option. The inability of the physical sciences to mark out the class of the politically corrupt constitutes evidence of the limitations of those sciences, not evidence that corrupt politicians do not exist. Physicalists then face a predicament. They can in good conscience neither countenance nor repudiate anthropic entities.

Recently, several philosophers have finessed the issue by adopting a hybrid that grafts nonrealism about anthropic matters onto scientific realism. The physical order is as it is, they contend, whatever we may think about it; moral, social, and political matters are what they are because of the ways we think about them and ourselves in relation to them. Scientific categories are discovered; intentional and social categories are contrived. Once we have constructed the category and formulated criteria for its application, there is such a thing as political corruption. But the scope of that category is severely limited in ways that natural kinds are not. There may, for example, be no saying whether stuffing the ballot box in an uncontested election is politically corrupt. This view underlies Bernard Williams's ethics,[3]

---

[2] The notion of explanatory adequacy here is roughly Hempel's. See Carl G. Hempel, "Aspects of Scientific Explanation," in *Aspects of Scientific Explanation and Other Essays* (New York: Macmillan, 1965), 331–496. Given a statement of the initial conditions and covering laws, the consequence—a description of the event to be explained—follows, or follows with a sufficiently high probability. It is worth noting that something could satisfy this criterion and be utterly unintelligible. A Hempelian explanation of the Watergate conspiracy that was cast entirely in the language of physics would be so long and so complicated that it is unlikely that anyone could understand it.

[3] Bernard Williams, *Ethics and the Limits of Philosophy* (Cambridge: Harvard University Press, 1985).

Allan Gibbard's moral psychology,[4] and Ian Hacking's social philosophy.[5]

Such a hybrid accords with weak supervenience. One sort of entity weakly supervenes on another just in case every entity of the one sort is an entity (or complex of entities) of the other, but there is no systematic correlation between the regularities the two sorts of entities display. Voting for president, for example, weakly supervenes on physical processes, since every casting of a ballot is a physical event, but the laws of physics cannot explain why U.S. presidential elections occur once every four years. If anthropic matters all weakly supervene on physical ones, then whatever is anthropic is physical, but physical regularities do not correlate with, hence physical laws do not explain, regularities in the anthropic realm. Weak supervenience enables the hybrid to evade ontological commitment to nonphysical entities. But since the physical sciences are acknowledged to be powerless to explain, predict, or control anthropic events, the position is physicalist in only the most attenuated sense.

Adherents of the hybrid do not agree about its consequences. Williams espouses a form of relativism; Gibbard, expressivism; Hacking, a type of historicism. Rather than discussing their differences, though, I want to explore a vision that they share. I will argue that the hybrid is neither viable nor necessary. Its plausibility stems not just from the irreducibility of the anthropic to the physical, but from irreducibility combined with a largely unexamined commitment to realism about the physical realm. A thoroughgoing constructive nominalism, I will urge, makes better sense of the phenomena than the hybrid does.

My discussion focuses on Hacking's work, which is a clear, concise, and forceful presentation of the hybrid. Because I will be criticizing Hacking's position, I want to begin with a disclaimer. The case studies Hacking develops are enormously valuable. They detail the social forces that were at work and the social interests that were served by introducing modes of classification

---

[4] Allan Gibbard, *Wise Choices, Apt Feelings* (Cambridge: Harvard University Press, 1990).
[5] Ian Hacking, "Making Up People," in *Reconstructing Individualism*, ed. T. C. Heller (Stanford: Stanford University Press, 1986), 222–236.

that made people into homosexuals, split personalities, and abused children. My quarrel is not so much with the picture as with the frame.[6]

Hacking contends that we ought to be realists about natural kinds and nominalists about social kinds. Only realism, he believes, can account for the success of science. If electrons exist and display the uniformities our theories say they do, there is no difficulty explaining the construction and deployment of experimental devices such as polarized electron guns.[7] But if the class of electrons is a motley collection of disparate entities, the fact that scientists have constructed and used such devices is inexplicable. Social kinds pretty clearly do not divide reality at antecedently existing joints. Nature cannot mark out the class of corrupt politicians independently of our political institutions and values. For the institutions we design and the values they embody determine what counts as corruption and what counts as political. That being so, Hacking opts for nominalism about the anthropic realm.

If the world actually divides into natural kinds whose constitution accounts for things behaving as they do, the success of a science that draws its lines elsewhere is a coincidence of enormous proportions. So for a metaphysical realist, the pull toward scientific realism is strong. It is stronger still for someone who thinks members of nominal kinds share nothing but their label, as Hacking says traditional nominalists do.[8] If, for example, the only thing electrons have in common is that they are called electrons, the effectiveness of polarized electron guns is a miracle. Scientists could no more expect to generate evidence of neutral bosons by marshalling electrons than they can expect to do the job by marshalling entities denoted by terms beginning with the letter e. If metaphysical realism is true and nominal kinds are arbitrary, nominalism about scientific kinds is implausible.

[6] Elsewhere, Hacking seems less committed to the frame. See, for example, Ian Hacking, "Worldmaking by Kindmaking: Child Abuse for Example," *How Classification Works*, ed. Mary Douglas and David Hull (Edinburgh: Edinburgh University Press, 1992), 180–238.

[7] Ian Hacking, *Representing and Intervening* (Cambridge: Cambridge University Press, 1983), 262–275.

[8] Hacking, "Making Up People," 227.

So is nominalism about anthropic kinds. If members of nominal kinds have nothing in common except that they are called by the same name, it is hard to see why one should be a nominalist about anything. Indeed, Hacking's case studies of anthropic concepts forcefully tell against such a nominalism.

One classification he has studied extensively is child abuse. A variety of behaviors qualify as child abuse: physical battering, sexual abuse, psychological traumatizing. If the class is arbitrary, it is fruitless to ask what else it might include. It *might* include anything—saxophones? charmed quarks? There is simply no saying. But the question is obviously not fruitless. Indeed, Hacking seems barely able to control his fury over the refusal of the "experts"—physicians, social workers, legislators, and the like—to recognize that child prostitution deserves to be recognized as child abuse.[9] The class is not the least bit arbitrary. It consists of terrible things that adults do to children. That we cannot provide a physicalistically acceptable criterion that marks out just those things makes no difference.

Hacking gives what he calls a "dynamic nominalist" construal of anthropic categories. On such a construal, "a kind of person came into being at the same time as the kind itself was being invented".[10] Until the concept is framed, there are no features shared by all and only the members of such a kind. This position is implausible. Even if the category "child abuse" is a recent addition to our conceptual repertoire, adults have intentionally done terrible things to children throughout history. Such acts always had something in common, even if our ancestors did not see fit to provide a label for their common feature.

The sort of nominalism I favor maintains that members of any collection are alike in some way or other. What labelling does is single out some likenesses as worthy of notice—worthy in light of our interests, enterprises, and purposes in categorizing. Any object belongs to infinitely many different classes and resembles the other members of each class it belongs to. Most resemblances, of course, are of no interest whatsoever. Doubtless, a spark plug, an

---

[9] Hacking, "Worldmaking by Kindmaking," 218–220.
[10] Hacking, "Making Up People," 228.

asteroid, and the Declaration of the Rights of Man have something in common. But it does not seem to be anything worth caring about. Some likenesses are less trivial. Membership in the class of miniskirts matters to couturiers, fashion plates, and enforcers of draconian dress codes. Membership in Princeton's incoming class, the United Mine Workers, the Royal Canadian Mounted Police matters a great deal to those who vie for admission.

Nominal kinds are not arbitrary then, but they are interest relative. We mark out certain classes and label them because (we hope) doing so will serve our purposes. Because we care desperately about children, we recognize a class of actions whose common feature is that they are particularly destructive of children's well-being. Because we have at least a passing interest in sex appeal and style, we mark out the class of miniskirts.

Realists contend that not all kinds are interest-relative. Objects cluster into natural kinds regardless of our interest in them. Horses, Hacking insists, are horses, no matter what anyone thinks.[11] Still, one wants to ask, why aren't zebras horses? Except for the stripes, they look like horses. Striped cats, after all, are cats. A realist has to answer: "That's just the way it is. Some things are horses and some aren't; and you can't always tell by the shape of the beast." The nominalist can say something more helpful— namely, that differentiating between horses and zebras serves certain identifiable interests. One derives from the inability of horses and zebras to interbreed. Given biology's interest in lineage, the capacity to interbreed is an important characteristic. Another derives from the fact that zebras, unlike horses, evidently cannot be domesticated. Our ancestors had good reason to care about, hence good reason to mark, the distinction between animals that are potential beasts of burden and animals that are not. Unlike the realist, the nominalist can often explain the rationale behind the categories we employ. She can say, "This is what favors one scheme of organization over another".

She can also recognize, as realists evidently cannot, that there are viable alternatives to the categories we actually use. Classifi-

[11] Ibid., 228–229.

cation of celestial bodies is for Hacking and other realists, a matter of getting at what is already there.[12] Plainly there are differences between Saturn, Titan, and Halley's Comet. There are nontrivial similarities as well. All are natural satellites. All travel in elliptical orbits. The main difference between moons and planets lies in the constitution of the objects they orbit; and the main difference between planets and comets lies in the eccentricity of their orbits. So where the realist takes the critical question to concern the reality of the similarities and differences, the nominalist takes it to be the importance of them. Had we introduced a single category applying to all and only celestial orbiters, moons, planets, and comets would have been the same kind of thing. It is not obvious that astronomy would have been worse off as a result.

In Chapter 1, I argued that the conduct of scientific inquiry affords no support for realism. Science seeks theories that satisfy its observational and theoretical desiderata—theories, that is, that are empirically adequate, simple, fruitful, explanatory, and so on. It imposes on its domain a scheme of organization that it takes to be conducive to framing such theories. Theories with different objectives may have reason to draw different lines. But whether or not the kinds it recognizes belong to a natural elite is a matter of indifference.

Still, Hacking asks, if nominalism is true, how could we have come up with categories that match the world so well? This is not so difficult, if we take the nominalist at her word. Entities always have something in common. So any system of categories matches the world somehow. An effective system does more. It invests the world with an order conducive to our ends. Such an order is not arbitrary. Nor do we invent schemes of classification ex nihilo. We begin with a conception of a domain, and a set of interests. Perhaps we also have some sense of why our current conception of things fails to answer to our interests. We then modify our categories to accommodate our interests and our interests to conform to our categories. Typically, a mutual adjustment of means and ends takes place. The belief that planetary orbits must be geometrically perfect drops out. If the glory of God is to be up-

---

[12] Ibid., 229.

held, it must be in some other way. For the ideal of perfect orbits does not mesh with the rest of astronomy. Similarly, questions of moral character are split off from questions of tubercular infection.[13] Doubtless there are morally weak people. Doubtless some of them contract tuberculosis. But having a so-called consumptive personality does not correlate with, much less explain or predict, susceptibility to the disease. So given the interests of medicine, the moral failings that make up a consumptive personality are irrelevant. Since the stereotype of the morally weak consumptive personality proves unhelpful for predicting, preventing, treating, and/or explaining tuberculosis, medicine has no use for it.

Hacking maintains that it is just a brute fact that tuberculosis is caused by microbes. He is right. But it is an equally brute fact that suicides are common in Scandinavia. And *suicide*, he contends, is a nominal kind. Whether the categories are natural or anthropic, once they are in place, the facts are equally brute. Even so, nominalism about natural kinds faces an objection:

> The constructive nominalist contends that the world does not divide itself into natural kinds. We make kinds and thereby make things into instances of them. It follows that *if* there were no concept of a dinosaur, there would be no dinosaurs. So it follows that *when* there was no concept of a dinosaur, there were no dinosaurs. In the Mesozoic era there was no concept of a dinosaur, for no one alive then had the intelligence to frame the concept. Therefore, in the Mesozoic era, there were no dinosaurs. There certainly have been no dinosaurs lately. So if there were no dinosaurs in prehistoric times, there were never any dinosaurs. Where did all the fossils come from?

If the challenge is sound, constructive nominalism about natural kinds is doomed. Luckily, it is not. Constructive nominalism is committed to the counterfactual claim that if there were no concept of a dinosaur, there would be no dinosaurs.[14] It is not committed to the historical claim that when there was no concept of a

---

[13] Ibid., 227.

[14] For a nominalist construal of counterfactuals, see Catherine Z. Elgin, *With Reference to Reference* (Indianapolis: Hackett, 1983), 50–54.

dinosaur, there were no dinosaurs. For once it is introduced, the concept of a dinosaur applies to all things, be they past, present, or future, that satisfy its criteria.

Concepts draw lines, and thereby distinguish one thing or kind from another. The concept of a dinosaur delineates criteria of identity for the beasts, specifying what it takes to be a dinosaur, what changes something can undergo and remain a dinosaur, what changes a species can undergo and remain a species of dinosaur, and so on. Without such criteria, it is indeterminate whether an iguanadon, or a pterondon, or even a hamster is a dinosaur. But once the criteria are fixed, they apply timelessly. The concept of a dinosaur is instantiated by all things that satisfy them, even if those things antedate the introduction of the category.

So let's sharpen the challenge:

> If no one had framed the concept of a dinosaur, there would have been no dinosaurs.

In this form, the challenge is ambiguous. On one reading it is true; on the other, false. If the term "dinosaur" is construed rigidly, the claim is obviously false. Given the reference that the term "dinosaur" actually has, there would have been dinosaurs, whether or not anyone so conceived of them. If "dinosaur" is construed nonrigidly, the challenge asks us to feign residence in a world where thinkers never framed the concept of a dinosaur. In such a world, if there were paleontologists, they would employ other concepts to make sense of their domain. But they would not group together all and only the animals we call dinosaurs. In that world, there would not have been dinosaurs.[15] We can read the

---

[15] A term (or concept) designates rigidly just in case it retains its actual reference in the counterfactual situation under discussion. If the counterfactual context is one in which the reference of the term can diverge from its actual reference, it designates nonrigidly. See Saul Kripke, "Naming and Necessity," in *Semantics of Natural Language*, ed. Donald Davidson and Gilbert Harman (Dordrecht: Reidel, 1972), 269–270. Unlike Kripke, I take the rigid/nonrigid distinction to be pragmatic, not metaphysical. One can, I think, contrive contexts in which even the most unnatural of kinds designates rigidly. For example, one might pose the examination question: "Would Kant have agreed that the emeralds in the evidence class are grue?" even though "grue" is an unnatural kind par excellence.

challenge either way. But unless we insist on giving it both readings simultaneously, it is toothless. The claim that scientists contrive natural kinds is neither paradoxical nor self-defeating. Natural scientists, like social scientists, humanists, journalists, and everyone else, devise schemes of categories that organize their domains in ways that they think will facilitate the advancement of the sort of understanding they seek. Natural kinds, on this usage, are just the kinds of interest to natural science. They are as nominal as any other kinds.

Hacking, however, considers the challenges successful. Natural kinds, he believes, cannot be nominal; for if they were, dinosaurs never existed. Anthropic kinds, he contends, being nominal, are severely restricted in scope. We make people into homosexuals, perverts, suicides, or consumptives, by contriving the categories for so describing them. If the categories were unavailable, Hacking concludes, people would not be homosexuals, perverts, suicides, or consumptives; and when the categories were unavailable, people were not homosexuals, perverts, suicides, or consumptives. Because the concept of a homosexual was not available before the end of the nineteenth century, for example, there were no homosexuals before that time.[16] But the inference is unsound, whether it pertains to dinosaurs or homosexuals. Once a concept has been formulated, it can designate rigidly and apply timelessly. The criterion for being a homosexual is, let us say, having or desiring sexual relations exclusively with members of one's own sex. Because prior to the end of the nineteenth century people satisfied this criterion, prior to the end of the nineteenth century there were homosexuals. The absence of the category from their conceptual repertoire means only that they could not think of themselves and each other as such.

We often describe historical figures, institutions, and events in ways the protagonists did not. And to good effect. Such descriptions enable us to discern features and patterns in the play of events that were not evident to the agents. We invest the world with an order its denizens did not recognize. Categories devised by contemporary economics yield insight into feudal institutions.

[16] Hacking, "Making Up People," 229.

Concepts like "mass hysteria" enrich our resources for understanding witch hunts. This is not to say that the agents' conceptions of their lives are inevitably wrong. They may well be right. But they are incomplete. There are indefinitely many ways to describe anything, and the agents have no monopoly on them.

Hacking's hybrid is historicist. Relativists often take a similar tack, with culture rather than history restricting the scope of anthropic concepts. Thus, many relativists contend that social and moral categories are applicable only within the societies that endorse them.[17] If the absence of the category of homosexuality in the eighteenth-century European conceptual repertoire entails that there were *then* no homosexuals, then the absence of that category in the conceptual repertoire of some distant contemporary culture should entail that there are *there* no homosexuals.

The real issue, of course, is not whether there are homosexuals at various times and places, but whether it is methodologically appropriate to classify people, actions, and institutions in ways that are alien to the agents' self-images. The social sciences regularly do so. The natural sciences and the humanities do too. Indeed some accurate and informative descriptions are all but absent from self-ascriptions. This does not discredit them. To explain an agent's behavior in terms of his unconscious motivation, repressed desires, or adrenal output is to put a gloss on it that he ordinarily cannot. It is a gloss that is often informative, even if he is entirely unaware of the categories in which it is framed. To label social movements pre-revolutionary is not possible until after the revolution has taken place. Yet without such a concept, it may be difficult to trace the trajectory of events that led to the revolution. And although religious adherents rarely think of their sincere religious beliefs and activities in terms of their contributions to the preservation of the status quo, appreciation of the stabilizing role religion plays may contribute significantly

---

[17] See Bernard Williams, "The Truth in Relativism," in *Moral Luck* (Cambridge: Cambridge University Press, 1981), 132–143; Alasdair MacIntyre, "Relativism, Power, and Philosophy," in *Relativism: Interpretation and Confrontation*, ed. Michael Krausz (Notre Dame: University of Notre Dame Press, 1989), 182–204.

to our understanding of a particular culture. A variety of structural and functional descriptions of political, cultural, and religious practices that sociologists and anthropologists employ are neither used or acknowledged by participants in those practices. Our understanding of the anthropic realm would be impoverished if they were unavailable to social scientists.

Hacking anchors his historicism in Elizabeth Anscombe's contention that an action is an action only under a description.[18] This we need not deny. For it does not follow that an action is or is merely the action that its agent describes it as. If categories apply retrospectively and cross-culturally, others may see behavior as actions that the agents do not recognize as such, and see behavior as different actions from those the agents take themselves to be performing. For that amounts to seeing that behavior instantiates certain action-descriptions that the agents do not acknowledge. We can say, though the authors of *Genesis* could not, that Cain's slaying Abel was an act of sibling rivalry. Any event has indefinitely many right descriptions. And human behavior is apt to have action-descriptions that are unavailable for self-ascription. Gavrilo Princip pulled a trigger, shot a gun, killed a man, assassinated a public figure, rid the world of a royal parasite, initiated the chain of events that led to World War I. He set out to do the first five. It is exceedingly unlikely that he set out to do the sixth. Still, he did it. And, arguably, that is the only historically noteworthy thing that he did.

Hacking contends that the historical record affords support for dynamic nominalism. If social categories apply retrospectively, he challenges, why is there no evidence of split personalities before the category was invented?[19] One possibility is that the category is in fact empty. Then there are no genuine instances, only misdiagnoses. But suppose we leave that possibility aside. Another possibility is that prior to the invention of the category, the significance of and relations among particular features of mental life were not recognized, hence not recorded. The category imposes a pattern on psychological phenomena, highlighting fea-

---

[18] Elizabeth Anscombe, *Intention* (Ithaca: Cornell University Press, 1957), 37–47.
[19] Hacking, "Making Up People," 223–224.

tures and connections we might otherwise overlook. Even if people had all the characteristics that are now considered symptomatic of a split personality, prior to the introduction of the concept there may have been no reason to construe them as symptoms, or as symptoms of the same condition. So their occurrence might easily have been ignored, downplayed, or ascribed to a range of divergent causes. In the absence of the category and of the understanding of the domain that the category contributes to, there was no reason to see the various manifestations as symptoms that fit together to form a psychologically significant syndrome.

We should expect to find the same phenomenon in other branches of medicine. A variety of seemingly unrelated characteristics are symptoms of toxemia—swollen fingers, protein in the urine, elevated blood pressure, seizures, and so on. Prior to the introduction of the concept of toxemia, and the initial steps toward understanding the disease, there was little reason to see such apparently heterogeneous symptoms as forming a pattern, hence little reason to record them all in the medical record. If you go to the doctor complaining of a headache, you are unlikely even to mention that your rings feel tight. And even if you happen to mention it, in the absence of a clinical picture connecting the two symptoms, the physician is not apt to consider it to have any bearing on your original complaint. If this is right, the history of medicine would reveal little evidence of a disease with multiple, seemingly unconnected symptoms before the disease concept was introduced. One might quibble over whether "disease" is itself an anthropic term, hence one that is subject to Hacking's historicism. However that issue is settled, "toxemia" denotes a physical condition that can and did cause deaths long before condition was named. If we consider how and why factors come to be mentioned in or excluded from medical records, the support Hacking finds for his historicism wanes.

The interest-relativity of nominal kinds leads some philosophers to look on them askance. One reason derives directly from realism. If there are more robust kinds that do not depend on us, the standing of nominal kinds seems somewhat tenuous. Another related reason is that nominal kinds are contrived. In Hacking's words, we "make up" people by creating categories for classifying them. A critical question is whether the categories we contrive are artificial or artifactual, whether our making is making *up* or making *into*. We make up fables. They are not true; for nothing in the world answers to them. If we make up kinds—"political corruption", "split personality", "comet", "horse"—nothing in the world (really) answers to them either. But we also *make* rags *into* rugs, meats and vegetables *into* stews, sentences *into* arguments. And if we are successful, the world contains rugs and stews and arguments as a result of our efforts. If the making of a nominal kind is a making into rather than a making up, the kind is an artifact, but is no less real for having been contrived.

Perhaps the source of concern is not contrivance per se, but the role of interests in the contrivance. Interests are supposed to be subjective. We happen to take an interest in other people's sex lives, but we might not have done so. We happen to care whether animals can be domesticated, but there was no necessity in that either. If categories like "heterosexual" and "homosexual", "horse" and "zebra" are a function of our interests, then they are tinged with subjectivity. That being so, they are not entirely objective.

In evaluating this line of thought, it is crucial that we consider what role interests actually play in the framing of categories. Again the parallel with artifacts is helpful. When I make a stew, I make it as I do because I have particular goals in mind. It's no accident that I add carrots and onions, and omit styrofoam and sawdust, for I want to produce a nourishing, palatable meal. I include or exclude garlic, coriander, and sage, to accommodate the tastes I ascribe to the diners. Still, my concoction has determinate, objective properties that are independent of the subjective interests, predilections, and objectives that influenced the making of it. Despite my good intentions and valiant efforts, the meal may be bland and nonnutritious. To my great surprise, the combination of spices I use may turn the gravy pink. Utterly unbeknownst to me, the stew may trigger Nina's nostalgia and Charlie's colitis. If the concoction has the consistency of concrete, it is not a stew. To say that an artifact is a product of intentional action, that it is designed and crafted to realize more or less specific goals, is not to say that it has all or only or indeed any of the features it was intended to have. Thinking you have made a nourishing, palatable stew does not make it so.

If the making that categorization effects is a making into rather than a making up, the results are artifacts with determinate, objective properties. We make people into homosexuals and heterosexuals by dividing the population on the basis of sexual preference. We make stegosauruses and tyrannosauruses into dinosaurs by grouping together prehistoric reptiles on the basis of anatomical similarities. We make certain rapists and batterers into child abusers on the basis of the special vulnerability of their victims. By drawing new lines then, we make things into members of the same kind.

The kinds we create are not artificial, nor are the similarities among their members subjective. There are determinate, objective facts about homosexuals, child abusers, and dinosaurs. And these facts may diverge considerably from our expectations. It may surprise us to learn that dinosaurs were highly adaptive, or that child abusers are often victims of abuse. We may discover that the categories we frame serve no useful purpose, or that they serve different purposes from those we had in mind. That "child abuser" has a determinate extension, for example, does not mean that it marks out a psychologically significant kind. We may discover that the category, though suitable for moral and legal purposes, is useless to psychiatry. Thinking it is a psychiatrically useful category does not make it so.

Both natural and anthropic kinds, I suggest, are contrived but determinate; interest-relative, but objective. In this they are like other human artifacts. Anthropic kinds differ from natural kinds not because the one sort are made, the other found, but because they serve different ends. That "political corruption" does not mark a class that is identifiable by the austere methods of the physical sciences should neither surprise nor distress us. It was designed to serve other purposes. And it serves those purposes admirably. It belongs to a system of anthropic kinds whose instantiation enables us to understand why Richard Nixon was forced to resign.

CHAPTER 11

The Relativity of Fact and the Objectivity of Value

Fact and value purport to be polar opposites: facts being absolute, material, objective, and impersonal; values relative, spiritual, subjective, and personal; facts being verifiable by the rigorous, austere methods of science; values being subject to no such assessment. The facts, they say, don't lie. So every factual disagreement has a determinate resolution. Whether barium is heavier than plutonium is a question of fact; and whatever the answer, there are no two ways about it. Values, if they don't precisely lie, are thought perhaps to distort. So evaluative disputes may be genuinely irresolvable. Whether, for example, a van Gogh is better than a Vermeer might just be a matter of opinion. And on matters like these, everyone is entitled to his own opinion. Such is the prevailing stereotype.

That stereotype ought to be rejected; for it stifles our understanding of both fact and value. Far from being poles apart, the two inextricably intertwine: the demarcation of facts rests squarely on considerations of value; and evaluations are infused with considerations of fact. So factual judgments are not objective unless value judgments are; and value judgments are not relative unless factual judgments are. I want to suggest that tenable judgments of both kinds are at once relative and objective.[1]

---

[1] Ruth Anna Putnam argues for a similar thesis in "Creating Facts and Values," *Philosophy* 60 (1985): 187–204.

First, let us look at the facts. When we proclaim their independence from and indifference to human concerns, we forget that we are the ones who set and enforce the standards for what counts as a fact. We stipulate: "a thing cannot both be and not be"; or "no entity without identity"; or "whatever is is physical". In effect we decree that whatever fails to satisfy our standards hasn't got what it takes to be a fact.

At the same time, we arrange for our standards to be met. We construct systems of categories that settle the conditions on the individuation of entities and their classification into kinds. Thus, for example, we devise a biological taxonomy according to which a dachshund is the same kind of thing as a Doberman, but a horse is a different kind of thing from a zebra.

For all their clarity, scientific examples may mislead. We are apt to think that constructing a biological taxonomy is simply a matter of introducing terminology for what is already the case. Then prior to our categorization, dachshunds and Dobermans were already alike; horses and zebras, already different. The problem is that any two things are alike in some respects and different in others. So likeness alone is powerless to settle matters of categorization. In classing dachshunds and Dobermans together, horses and zebras apart, we distinguish important from unimportant similarities. That is, we make a value judgment.

The selection of significant likenesses and differences is not, in general, whimsical. It is grounded in an appreciation of why a particular classificatory scheme is wanted; and this, in turn, depends on what we already believe about the subject at hand. If our goal is to understand heredity, for example, it is reasonable to group together animals that interbreed. Then despite their obvious differences, dachshunds and Dobermans belong together; and despite their blatant similarities, horses and zebras belong apart.

More general considerations come into play as well. If our system is to serve the interests of science, the cognitive values and priorities of science must be upheld. Membership in its kinds should be determinate and epistemically accessible. There should be no ambiguity and no (irresolvable) uncertainty about an individual's membership in a kind. The classification should be conducive to the formulation and testing of elegant, simple, fruitful generalizations, and should perhaps mesh with other scientific classifications of the same and adjacent domains. In constructing a system of categories suitable for science then, we make factual judgments about what the values of science are, and how they can be realized.

Science streamlines its categories in hopes of achieving exceptionless, predictive, quantitative laws. Narrative has quite different ends in view, being concerned with the particular, the exceptional, the unique. So schemes suited to narrative enterprises exhibit different features from those suited to science. Scientific vices—ambiguity, imprecision, immeasurability, and indeterminacy—are often narrative virtues.[2] The complex characterization of the emotional life that we find, for example, in the novels of Henry James requires a baroque conceptual scheme whose involuted categories intersect in intricate and subtle ways. Equally complex categories may be required to achieve the sort of understanding that biographers, historians, psychoanalysts, and serious gossips strive to achieve.

A category scheme provides the resources for stating various truths and falsehoods, for exhibiting particular patterns and discrepancies, for drawing specific distinctions, for demarcating conceptual boundaries. Purposes, values, and priorities are integral to the design. They constitute the basis for organizing the domain in one way rather than another. The acceptability of any particular scheme depends on the truths it enables us to state, the methods it permits us to employ, the projects it furthers, and the values it promotes. Together, these constitute a system of thought. A failure of the components to mesh undermines the system, preventing it from doing what it ought to do.

We design category schemes with more or less specific purposes in mind and integrate into the scheme such values and priorities as we think will serve those purposes. But the values that our schemes realize are not always or only the ones we intend to produce. Some are simply mistakes; others, inadvertent hold-

[2] Israel Scheffler, *Beyond the Letter* (London: Routledge and Kegan Paul, 1979), 6–7.

overs from prior systems; yet others, unintended byproducts of features we intentionally include. When pregnancy and aging are classified as medical conditions, they come to be considered, and treated as diseases or disabilities—as deviations from a state of health. If Marx is right, the values of the ruling class are invisibly embedded in the social and economic categories of a society. And my students are convinced that a fundamental truth is revealed by the fact that witchcraft comes just after philosophy in the Library of Congress classification system.

As a first approximation, facts are what answer to true sentences. And different systems produce different truths. It is a truth of physics, not of botany, that copper is lighter than zinc. This alone does not lead to relativity, for such systems may complement one another, or be indifferent to one another. Relativity emerges when systems clash—when what is true according to one system is false according to another. Evolutionary taxonomy so groups animals that crocodiles and lizards are close relatives; crocodiles and birds, distant ones. Cladistic classification shows crocodiles and birds to be close; crocodiles and lizards distant. Each system divulges some affinities among animals and obscures others: neither invalidates the other. So whether it is a fact that crocodiles and lizards are closely related depends on a choice of system. According to one system, any violation of the law is a crime; according to another, only serious violations—felonies—are crimes. So whether spitting on the sidewalk is a crime depends on which system is in use. According to one medical classification, health is the absence of disease; according to another, health is the absence of disease or disability. So whether a congenital defect renders a person unhealthy depends on which system is in effect. A single domain can be organized in a multitude of ways, and different schematizations may employ a single vocabulary. So under one schematization a given sentence—say, "Spitting on the sidewalk is a crime"—comes out true; under another, it comes out false. Truth then is relative to the system in effect.

Still, facts are objective. For once the system is in place, there is no room for negotiation. Events that are simultaneous relative to one frame of reference are successive relative to another. But it is determinate for each frame of reference whether given events are successive or simultaneous. Similarly, although some psychologistic systems consider neuroses to be mental illnesses and others do not, once a system is chosen, there is a fact of the matter as to whether a compulsive handwasher is mentally ill.

Such objectivity might seem spurious, if we can switch frameworks at will. What is true according to one framework is false according to another. So can't we just choose our facts to fit our fancy? There are at least two reasons why we cannot. The first is that rightness requires more than truth.[3] We need to employ an appropriate framework, one that yields the right facts. The fact that someone went to Choate, for example, neither qualifies nor disqualifies him for a federal judgeship. So a classification of candidates according to their secondary schools is inappropriate, even if it would enable us to choose the candidate we want. Correctness requires that the facts we appeal to be relevant. Psychoanalytic categories are powerless to settle the issue of criminal insanity because they mark the wrong distinctions. People who cannot be held criminally liable for their actions are supposed to be, in some important respect, different from the rest of us. And the categories in question reveal no difference; for they characterize everyone's behavior in terms of motives and desires the agent can neither acknowledge nor control. So the facts that psychoanalytic theory reveals do not suit the purposes of the criminal court; they do not discriminate the class of the criminally insane. Rightness of categorization thus depends on suitability to a purpose. An aspiring lepidopterist whose collection consists of larvae seems to have missed the point. Lepidopterists concentrate on mature forms—they collect butterflies, not caterpillars. Although biologists class butterflies and caterpillars together, butterfly collectors do not. Rightness here requires fit with past practice. The fellow fails as a lepidopterist because he employs radically nontraditional categories in selecting specimens for his collection.

Moreover, even though we construct the categories that fix the facts, we cannot construct whatever we want. If we take the notion of construction seriously, this will come as no surprise.

[3] Nelson Goodman, *Ways of Worldmaking* (Indianapolis: Hackett, 1978), 109–140.

Although we make all manner of inventions, we cannot make a nonfattening Sacher Torte, a solar-powered subway, or a perpetual-motion machine. And although we design programs that endow computers with amazing abilities, we cannot get a computer to translate a natural language, compute the last digit in the decimal expansion of π, or consistently beat a grand master at chess.

Some of these incapacities are irremediable; others will eventually be overcome. My point in mentioning them is to emphasize that construction is something we do; and we cannot do everything we want. Our capacities are limited; and our aspirations often interfere with one another. So there is no reason to think that we can convert every fantasy into fact by designing a suitable system. Plainly, we cannot.

In constructing a political system, for example, we would like to maximize both personal liberty and public safety. We would like, that is, to arrange for as many actions as possible to fall under the predicate "free to . . ." and as many harms as possible to fall under the predicate "safe from . . .". But we cannot maximize both at once. The cost of security is a loss of liberty; and the cost of liberty, a risk of harm. With the freedom to carry a gun comes the danger of getting shot. So we have to trade the values of liberty and safety off against each other to arrive at a system that achieves an acceptable level of both.

In constructing a physicalistic system, we would like all the magnitudes of elementary particles to be at once determinate and epistemically accessible. But this is out of the question. For although we can measure either the position or the momentum of an electron, we cannot measure both at the same time.

In building a system of thought, we begin with a provisional scaffolding made of the (relevant) beliefs we already hold, the aims of the project we are embarked on, the liberties and constraints we consider the system subject to, and the values and priorities we seek to uphold. We suspend judgment on matters in dispute. The scaffolding is not expected to stand by itself. We anticipate having to augment and revise it significantly before we have an acceptable system. Our initial judgments are not comprehensive; they are apt to be jointly untenable; they may fail to serve the purposes to which they are being put, or to realize the values we want to respect. So our scaffolding has to be supplemented and (in part) reconstructed before it will serve.

The considered judgments that tether today's theory are the fruits of yesterday's theorizing. They are not held true come what may, but accorded a degree of initial credibility because previous inquiry sanctioned them. They are not irrevisable, but they are our current best guesses about the matter at hand. So they possess a certain inertia. We need a good reason to give them up.[4]

System building is dialectical. We mould specific judgments to accepted generalizations, and generalizations to specific judgments. We weigh considerations of value against antecedent judgments of fact. Having a (partial) biological taxonomy that enables us to form the generalization "like comes from like"—that is, progeny belong to the same biological kind as their parents— we have reason to extend the system so as to classify butterflies and caterpillars as the same kind of thing. Rather than invoke a more superficial similarity and violate an elegant generalization, we plump for the generalization and overlook obvious differences.

Justification is holistic. Support for a conclusion comes not from a single line of argument, but from a host of considerations of varying degrees of strength and relevance. What justifies the categories we construct is the cognitive and practical utility of the truths they enable us to formulate, the elegance and informativeness of the accounts they engender, the value of the ends they promote. We engage in system building, when we find the resources at hand inadequate.[5] We have projects they do not serve, questions they do not answer, values they do not realize. Something new is required. But a measure of the adequacy of a novelty is its fit with what we think we already know. If the finding is at all surprising, the background of accepted beliefs is apt to require modification to make room for it; and the finding may require revision to be fitted into place. A process

---

[4] Nelson Goodman, "Sense and Certainty," in *Problems and Projects* (Indianapolis: Hackett, 1972), 60–68.

[5] But not only then. We may attempt to modify a working system out of curiosity—to see how it works and whether it can be made to work better.

of delicate adjustments occurs, its goal being a system in wide reflective equilibrium.[6]

Considerations of cognitive value come into play in deciding what modifications to attempt. Since science places a premium on repeatable results, an observation that cannot be reproduced is given short shrift, while one that is readily repeated may be weighted so heavily that it can undermine a substantial body of theory. A legal system that relies on juries consisting of ordinary citizens is unlikely to favor the introduction of distinctions so recondite as to be incomprehensible to the general public.

To go from a motley collection of convictions to a system of considered judgments in reflective equilibrium requires balancing competing claims against one another. There are likely to be several ways to achieve an acceptable balance. One system might, for example, sacrifice scope to achieve precision; another, trade precision for scope. Neither invalidates the other. Nor is there any reason to believe that a uniquely best system will emerge in the long run.

To accommodate the impossibility of ascertaining both the position and the momentum of an electron, drastic revisions are required in our views about physics. But which ones? A number of alternatives have been suggested. We might maintain that each electron has a determinate position and a determinate momentum at every instant, but admit that only one of these magnitudes can be known. In that case, science is committed to the existence of things that it cannot in principle discover. Or we might contend that the magnitudes are created in the process of measurement. Then an unmeasured particle has neither a position nor a momentum, and one that has a position lacks momentum, since the one measurement precludes the other. Physical magnitudes are then knowable because they are artifacts of our knowledge-gathering techniques. But from the behavior of particles in experimental situations, nothing follows about their behavior elsewhere. Yet a third option is to affirm that a particle has a position and affirm

[6] See Nelson Goodman, *Fact, Fiction, and Forecast* (Cambridge: Harvard University Press, 1984), 65–68; John Rawls, *A Theory of Justice* (Cambridge: Harvard University Press, 1971); Catherine Z. Elgin, *With Reference to Reference* (Indianapolis: Hackett, 1983), 183–193.

that it has a momentum, but deny that it has both a position and a momentum. In that case, however, we must alter our logic in such a way that the conjunction of individually true sentences is not always true. That science countenances nothing unverifiable, that experiments yield information about what occurs in nature, that logic is independent of matters of fact—such antecedently reasonable theses are shown by the findings of quantum mechanics to be at odds with one another. Substantial alterations are thus required to accommodate our theory of scientific knowledge to the data it seeks to explain. Although there are several ways of describing and explaining quantum phenomena, none does everything we want. Different accommodations retain different scientific desiderata. And deciding which one to accept involves deciding which features of science we value most, and which ones we are prepared, if reluctantly, to forego. "Unexamined electrons have no position" derives its status as fact from a judgment of value—the judgment that it is better to construe magnitudes as artifacts of measurement than to modify classical logic, or commit science to the truth of claims it is powerless to confirm, or to make any of the other available revisions needed to resolve the paradox.

Pluralism results. The same constellation of cognitive and practical objectives can sometimes be realized in different ways, and different constellations of cognitive and practical objectives are sometimes equally worthy of realization. A sentence that is right according to one acceptable system may be wrong according to another.

But it does not follow that every statement, method, or value, is right according to some acceptable system. Among the considered judgments that guide our theorizing are convictions that certain things—for instance, affirming a contradiction, ignoring the preponderance of legal or experimental evidence, or exterminating a race—are just wrong. Such convictions must be respected unless we find powerful reasons to revise them. There is no ground for thinking that such reasons are in the offing. It is not the case that anything goes.

Nor does it follow that systems can be evaluated only by standards that they acknowledge. An account that satisfies the standards it sets for itself might rightly be faulted for being blind to problems it ought to solve, for staking out a domain in which there are only trivial problems, for setting too low standards for itself. An inquiry that succeeds by its own lights may yet be in the dark.

So far, I have argued for the value-ladenness of facts. I developed a scientific example in some detail, because science is considered a bastion of objectivity. If scientific facts can be shown to be relative and value-laden, there is a strong *prima facie* case for saying that relativity and value-ladenness do not undermine objectivity. If the objectivity of normative claims is to be impugned, then it must be on other grounds.

I want to turn to questions of value. Not surprisingly, I contend that value judgments are vindicated in the same way as factual judgments. Indeed, normative and descriptive claims belong to the same systems of thought, and so stand or fall together. Still, some systems seem more heavily factual; others, more heavily evaluative. For now, I will concentrate on the latter.

In constructing a normative category scheme, as in constructing any other scheme, we are guided by our interests, purposes, and the problem at hand. Together these factors organize the domain so that certain considerations are brought to the fore. In reforming the zoning laws, for example, it is advisable to employ consequentialist categories. For we need the capacity to tell whether things would in fact improve if the building code were altered in one way or another. We need then the capacity to classify and to evaluate in terms of outcomes. If we are concerned with developing moral character, it may be advisable to use predicates that can be applied with reasonable accuracy in self-ascription; for the capacity for self-scrutiny is likely to be a valuable asset in moral development. For like cases to be treated alike, the evaluations yielded by a moral or legal system must be coherent, consistent with one another, and grounded in the relevant facts. Fairness and equity are demanded of such a system; arbitrariness and caprice are anathemas to it. So logical and evidential constraints are binding on evaluation as well as on description.

The problems we face and the constraints on their solution

often have their basis in the facts. Whether, for example, we ought to perform surgery to prolong the life of a severely defective newborn becomes a problem only when we acquire the medical resources to perform such surgery. Prior to the development of the medical techniques, the question was moot. There was no reason to require a moral code to provide an answer. So a moral problem arises in response to changes in the facts.

Our previously acceptable moral code may never have needed, and so never have developed, the refinements required to handle the new case. Unanticipated facts can thus put pressure on a system, by generating problems it cannot (but should) solve, yielding inconsistent evaluations, or producing counterintuitive verdicts. Values that do not ordinarily clash may do so in special circumstances. Typically the physician can both prolong the lives of her patients and alleviate their pain. But not always. So a moral system that simply says she ought to do both is inadequate. It does not tell her how to proceed when the realization of one value interferes with the realization of the other. Our values then need to be reconsidered. In the reconception, previously accepted conclusions are called into question, competing claims adjudicated, a new balance struck. Our goal again is a system of considered judgments in reflective equilibrium. Achieving that goal may involve drawing new evaluative and descriptive distinctions or erasing distinctions already drawn, reordering priorities or imposing new ones, reconceiving the relevant facts and values or recognizing new ones as relevant. We test the construction for accuracy by seeing whether it reflects (closely enough) the initially credible judgments we began with. We test it for adequacy by seeing whether it realizes our objectives in theorizing. An exact fit is neither needed nor wanted. We realize that the views we began with are incomplete, and suspect that they are flawed; and we recognize that our initial conception of our objectives is inchoate, and perhaps inconsistent. So we treat our starting points as touchstones which guide but do not determine the shape of our construction.

Here too, pluralism results; for the constraints on construction need not be uniquely satisfied. Where competing considerations are about equal in weight, different tradeoffs might reasonably be made, different balances struck. If any system satisfies our standards, several are apt to do so.

In childrearing, for example, we regularly have to balance concern for a child's welfare against the value of granting him autonomy. Responsible parents settle the matter differently, some allowing their children greater freedom, some less. A variety of combinations of permissions and prohibitions seem satisfactory, none being plainly preferable to the rest. It follows then that a single decision—say, to permit a child to play football—might be right or wrong depending on which acceptable system is in effect. Rightness is then relative to system.

But it does not follow that every act is right according to some acceptable system or other. It is irresponsible to permit a toddler to play with matches, and overprotective to forbid a teenager to cross the street. From the fact that several solutions are right, it does not follow that none is wrong. Some proposed resolutions to the conflict between welfare and autonomy are plainly out of bounds.

Nor does it follow that to be right according to some acceptable system is to be right *simpliciter*. Rightness further requires that the system invoked be appropriate in the circumstances. Although most freshmen philosophy papers would rightly be judged abysmal failures if evaluated according to the editorial standards of the *Journal of Philosophy*, those are clearly the wrong standards to use. To grade students fairly, we must employ standards appropriate to undergraduate work. (Then only some of their papers are abysmal failures.)

Can we rest satisfied with the prospect of multiple correct evaluations? Disconcertingly, the answer varies. If the systems that produce the several evaluations do not clash, there is no difficulty. We easily recognize that an accurate shot by an opposing player is good from one point of view (excellence in playing the game) and bad from another (our partisan interest that the opposition collapse into incompetence). And there is no need to decide whether it is a good or bad shot, all things considered.

In other cases, multiplicity of correct evaluations may be rendered harmless by a principle of tolerance. We can then say that what is right according to any acceptable system is right. Thus one parent's decision on how best to balance paternalist and libertarian considerations in childrearing does not carry with it the commitment that all parents who decide otherwise are wrong. And one physician's decision on how to balance the value of alleviating pain against the value of prolonging life does not carry with it the commitment that all physicians who strike a different balance are wrong.

Tolerance is an option because the prescriptions for action apply to numerically distinct cases. So long as parents decide only for their own children and both parents agree about that, they can recognize that other parents might reasonably decide the same matters somewhat differently. Pluralism does not lead to paralysis here because the assignment of responsibility is such that conflicting right answers are not brought to bear on a single case.

Tolerance seems not to be an option, however, when systems dictate antithetical responses to a single case. For we must inevitably do one thing or another. The problem becomes acute in socially coordinated activity. If the several parties in a joint venture employ clashing systems, their contributions are likely to cancel each other out, diminishing the prospect of success. Although nothing favors the convention of driving on the right side of the road over that of driving on the left, leaving the choice to the individual driver would be an invitation to mayhem. We need then to employ a single system, even if the selection among acceptable alternatives is ultimately arbitrary.

In such cases, then, we invoke a metasystematic principle of intolerance. Even if there are several ways of equilibrating our other concerns, we mandate that an acceptable equilibrium has not been reached until a single system is selected. The justification for this mandate is the recognition that unanimity or widespread agreement is itself a desideratum that is sometimes worth considerable sacrifice to achieve.

To be sure, an intolerant system remains vulnerable to criticism, revision and replacement by a better system. The argument for intolerance is simply that where divided allegiance leads to ineffectiveness, a single system must reign. Successors there can be, but no contemporaries.

In the cases I have spoken of so far, both tolerance and intolerance look like fairly easy options. We readily agree to be intolerant about rules of the road, not only because we appreciate the value of conformity in such matters, but also because we recognize that nothing important has to be given up to achieve conformity. It simply does not matter whether we drive on the left or on the right, so long as we all drive on the same side. And we readily tolerate a range of childrearing practices, because so long as certain broad constraints are somehow satisfied, small differences don't much matter. The difference between a 10 and a 10:30 PM curfew is unlikely to significantly affect a child's well being. In such cases we can agree, or respectfully agree to disagree, precisely because no deeply held convictions are violated in the process.

Sometimes, however, conflicts run deep. For example, the abortion problem arises because in an unwanted pregnancy, the value of personal autonomy clashes with the value of fetal life. Neither is trivial. So to achieve any resolution, a substantial good must be sacrificed. Each party to the dispute achieves equilibrium at a price the other is unwilling to pay: the one maintaining that even fetal life cannot compensate for the loss of liberty, the other maintaining that even liberty cannot compensate for the loss of fetal life. Nor, evidently, can the parties civilly agree to disagree. Each is convinced that the position of the other is fundamentally immoral. Both parties to the dispute can adduce powerful reasons to support their position. But neither has the resources to convince its opponents. Nor has anyone come up with a compromise that both sides can in good conscience accept.

The existence of such seemingly intractable problems might seem to support a subjective ethical relativism. Having found no objective way to resolve such dilemmas, we might conclude that all morality is relative to system, and the choice of a system is, in the end, subjective. Without denying the difficulty that such problems pose, I want to resist the slide into subjectivism. Our practice bears me out. Even in the face of widespread disagreement, we do not treat such issues as subjective. If we did, we would probably be more charitable to those holding opposing views. How do we proceed?

Sometimes we deny that the problem remains unsolved. We contend that one of the positions, although still sincerely held, has actually been discredited. The holdouts, we maintain, overlook some morally relevant features of the situation, or improperly weigh the relevant ones. This response may well be correct. Advocates of apartheid, however adamant, are just wrong. And they remain wrong even if they are too ignorant, biased, or stupid to recognize their error. So the failure of an argument to convince its opponents may be due to defects in their understanding or their character, not to weaknesses in the argument. This has its parallel in science. The inability of any argument to convince my accountant of the truth of the Heisenberg uncertainty principle does not discredit the objectivity of the principle; it discredits her claim to have mastered quantum mechanics.

Alternatively, we might concede that a question is unanswered, without concluding that it is unanswerable. We then take it to be an outstanding problem for the relevant field of inquiry. All fields have such problems. And if our current inability to solve the problem of origin of life does not impugn the objectivity of biology, our current inability to reconcile our attitudes toward pornography and freedom of speech should not impugn the objectivity of ethics. What such problems show is that work remains to be done. This is no surprise.

The objectivity of ethics does not ensure that we can answer every question. Neither does the objectivity of science. If a question is ill-conceived or just too hard, or if our attempts are wrongheaded or unlucky, the answer may forever elude us. That success is not guaranteed is just an epistemological fact of life.

Nor does objectivity ensure that every properly conceived question has a determinate answer. So perhaps nothing determines whether the young man whom Sartre describes ought to join the Resistance or stay home and care for his aged mother.[7] If the relevant considerations are in fact equally balanced, either alternative is as good (or as bad) as the other. The choice he faces then is subjective. But this does not make ethics subjective. For to say that personal predilections are involved in deciding among

---

[7] Jean-Paul Sartre, *Existentialism and Humanism* (London: Methuen, 1968), 35–37.

equally worthy alternatives is quite different from saying that personal predilections are what make the alternatives worthy. Subjective considerations function as tie breakers after the merit of the contenders has been certified by other means.

I have suggested that factual and evaluative sentences are justified in the same way. In both cases, acceptability of an individual sentence derives from its place in a system of considered judgments in reflective equilibrium. Since equilibrium is achieved by adjudication, several systems are apt to be adequate. But since they are the products of different tradeoffs, they are apt to disagree about the acceptability of individual sentences. So relativism follows from pluralism. Something that is right relative to one acceptable system may be wrong relative to another.

Still, the verdicts are objective. For the systems that validate them are themselves justified. The accuracy of such a system is attested by its ability to accommodate antecedent convictions and practices; its adequacy, by its ability to realize our objectives. Several applicable systems may possess these abilities; so several answers to a given question, or several courses of action may be right. But not every system possesses them; so not every answer or action is right. The pluralism and relativism I favor thus do not lead to the conclusion that anything goes. If many things are right, many more remain wrong.

CHAPTER 12

Postmodernism, Pluralism, and Pragmatism

The problem, it seems, is difference of opinion. More specifically, how should such differences be understood? It is unlikely that disagreements are more numerous or more profound today than they were in the past. But their occurrence seems more problematic. Our forebears believed that every disagreement had a resolution. Genuine, mind-independent facts were supposed to suffice to settle every controversy. To be sure, debates raged fiercely, even violently; for the facts that were supposed to decide things were elusive, and sometimes hotly contested. Thus, for example, the dispute between Galileo and the Church was at least in part a dispute over what sorts of facts count as astronomical evidence. The debates and disputes of our forebears may have been as bitter, and their resolution as remote as the debates and disputes of today. But they seem to have had a quite different cast. For they were conducted in absolutist terms. At most one of the parties to any disagreement could be right. This much, if no more, both sides conceded. The difficulty of the problem or the boneheadedness of one's opponent was adduced to explain failure to reach accord.

These days we are not so sure. The intellectual enterprises that were expected to secure knowledge and fix the facts have failed abysmally. We have found no incontrovertible canons of ethical or aesthetic evaluation, no infallible scientific methods, no clear and direct sources of religious insight, no modes of argument so persuasive as to be universally compelling. Even mathematical and logical knowledge is not all we had hoped it would be; proof procedures and the resources they rely on are contestable in principle and contested in fact.

What undermines our confidence is not only (or even, I think, mainly) philosophy's failure to discover incontrovertible truths. Even more disconcerting is humanity's failure *in practice* to achieve lasting accord. That the disputes rage as loudly and fiercely as ever, that we seem no closer to consensus despite the best efforts of our best people, may make us doubt that accord will ever be reached. The postmodern predicament is not the loss of absolutes—Truth, Goodness, Certainty, and the like. We never had them. It is, rather, the loss of the expectation and perhaps the hope of achieving them. What should we think?

We might continue to maintain that differences of opinion are genuine disagreements about what is in fact the case. Then there is exactly one correct answer to any question, and the problem is to discover that answer and intellectually compelling reasons to believe it. This leads to dogmatism or to despair. We cling tenaciously to our own convictions, denying the force of the arguments that tell against them, even if we have no satisfactory answers to those arguments. Or we face up to the history of failures to reach accord, and conclude that the absolutes we seek are beyond our ken. Our paltry efforts to achieve knowledge are futile, for the conditions on knowledge are too demanding. Descartes' demon wins.

Alternatively, we might conclude that where there is no hope of resolution, there is no real dispute. One option is as good as another; for there are no facts to decide between them. Anything goes. What results is subjectivism in place of epistemology, hedonism in place of ethics, sociology in place of criticism. We can keep score: tallying which contentions are actually believed, which actions are actually approved, which works of art are actually enjoyed. But we have no obligation to determine the warrant for beliefs, the merit of actions, the excellence of works of art. For there are no objective standards for making such evaluations. There are only opinions.

Vacillation between the absolute and the arbitrary stems from a failure to recognize the availability of an alternative. Either there is one right answer or there is not. That is obviously true. The error arises when we interpret the second disjunct as "or there is none". One way there can fail to be one right answer is that there is none. Another is that there are several. To say that a problem does not admit of a unique solution is not to say that it is unsolvable or that all proposed solutions are equally good. A math student asked to give the square root of 4 can correctly answer $+2$ or -2. But the fact that there are two correct answers does not entail that every answer is correct. She cannot hope to get credit if she answers 17. Likewise, a work like *Madam Bovary* admits of multiple correct interpretations. But not every interpretation is correct. The work cannot plausibly be construed as a commentary on the fall of the Roman Empire or as a story about a boy and his dog.

Once we admit that there may be multiple, equally good ways of understanding or doing things, differences of opinion take on a new guise. Not every difference of opinion is a disagreement, and not every disagreement has a resolution. From the fact that we differ, it does not follow that at least one of us is wrong. Nor does it follow that we are both right. Each alternative has to be considered on its merits. But how do we determine these?

We're back, it seems, where we started. Either there are shared standards for deciding such matters or there are not. If there are, they may seem to supply the absolutes we seek. The student is justified in saying either $+2$ or -2 is the square root of 4 because both answers satisfy the same standard: namely, the square root of x is the number that multiplied by itself yields x. If the standards that vindicate alternative construals are always common to those construals, pluralism looks innocuous. For the standards seem absolute, even if multiply satisfiable. But appearances can be misleading. Even shared standards are not always acceptable. Trial by dunking is not an acceptable standard for determining who is a witch, even if everyone thinks it is. Consensus is not the hallmark of correctness. Moreover, standards themselves are often in dispute. When, for example, physicists disagree about whether unexamined electrons have a determinate position or legal theorists disagree about whether a right to privacy is conferred by the U.S. Constitution, they often also disagree about what standards are to decide such matters. With or without consensus, a regress looms. Unless it can be blocked, there seems no way to rule any account untenable.

We need not embark on the regress, though. In deciding whether scientific, legal, or other sorts of standards are acceptable, we do not ordinarily appeal to ever more abstract principles that are supposed to subsume them. Usually, we look closer to home. Widely accepted standards are often called into question by the instances they are supposed to govern. The Russell paradox discredits naive set theory's criterion of membership by showing that it leads to contradiction. If sets can take anything as members, there is a set that contains all sets that do not contain themselves. It is then a member of itself if and only if it is not. The popularity of the criterion is powerless to defend it against the Russell set.

The grue paradox undermines a seemingly obvious and widely accepted standard of inductive support. If a generalization is inductively supported just in case all the objects in its evidence class instantiate it, "All emeralds are grue" has massive support. This is not contradictory; but it is highly counterintuitive. And if we retain the standard and accept the consequence, induction loses its point. For it gives us no reason to infer "All emeralds are green" rather than "All emeralds are grue", or any of the infinitely many alternatives that do not conflict with the evidence.

A case need not be paradoxical to call a standard into question. When grand juries refuse to indict people for mercy killings, they manifest dissatisfaction with legal standards that construe every taking of an innocent life as murder. They don't consider the standards paradoxical, just wrong. When connoisseurs balk at concluding that Vermeer's paintings are devoid of monetary value, they call into question the economic standard that takes the value of a thing to be the price a willing buyer will pay to a willing seller. For the reason there is no market in Vermeers is that no one who owns a Vermeer is willing to sell it at any price. That would seem to show that the paintings are immensely valuable, not valueless. Standards then are not always binding. They are subject to revision if they clash with verdicts we are loathe to reject.

Sometimes it is the fact of disagreement itself that gives us pause. The recognition that responsible, knowledgeable biologists believe that evolution proceeds by punctuated equilibrium may undermine the confidence of the ardent evolutionary gradualist. The recognition that verdicts that seem to issue from a constitutional right to privacy can be readily accommodated by appeal to other, less controversial rights may cause misgivings in legal theorists committed to a constitutional right to privacy.

Not every disagreement is equally unsettling. Only if an alternative is grounded in the right sort of reasons should it raise doubts. The gradualist is justifiably concerned by the alternative offered by adherents of punctuated equilibrium, for that alternative respects the canons of biological science. She has no reason to take the biological claims of Christian fundamentalists seriously, since they fly in the face of the evidence the fossil record presents.

When we start talking about the right sorts of reasons though, the prospect of the regress recurs. What makes reasons the right sort? The pull between the absolute and the arbitrary exerts itself anew: either the nature of things somehow mandates that biologists ought to attend to the fossil record and ignore religious texts, or their taking the fossil record rather than the Bible as their source of evidence is capricious, or merely sociological. Neither is remotely plausible.

Rather, as I argued in Chapter 11, we justify our theories by the method of reflective equilibrium. Locke to the contrary notwithstanding, we do not start with an empty slate. We begin any inquiry with a host of beliefs, standards, methods, and values that we are inclined to accept and consider relevant to the subject at hand. These are our working hypotheses. They are apt to be inadequate. They may be incomplete, or mutually inconsistent, or entail consequences that we cannot on reflection endorse. If so, we augment and revise them until we arrive at a constellation of commitments that we consider acceptable. The elements of such a constellation must be reasonable in light of one another, and the constellation as a whole must be reasonable in light of our antecedent commitments. This does not mean that it must incorporate all the commitments we started with. We expect working hypotheses to undergo revision. Still, the account we arrive at should be able to explain in retrospect why the commitments we began with were successful to the extent that they were.

The commitments in question are not all beliefs about the subject matter. We bring to an inquiry methodological commitments, techniques, criteria, and objectives. All provide grist for the mill. Like beliefs, they are subject to revision and rejection in the process of constructing a tenable system of thought. Even if we seek a biology that accords with a literal interpretation of the Bible, there is no guarantee that we can contrive one without violating other commitments the science is more strongly invested in.

The various components of a system of thought constrain one another. What justifies the biologist in taking the fossil record to afford the right sort of evidence is the success of previous biological theories that relied on that record. And what justifies her in taking recent biological theories to be reasonably successful is the scope and depth of the understanding of living things that they afford. Her justification has other strands as well: the success of geology in explaining how fossils are formed; the success of genetics in explaining the mechanisms of evolution; the success of philosophy of science in vindicating the methods of empirical science, and so on.

Standards of acceptability are themselves products of reflective equilibrium. In contriving such standards we begin with whatever standards we actually endorse, and with cases we actually accept. These, we are apt to discover, fail to mesh, or to extend as far as we think they should, or vindicate the accounts we consider acceptable. We cannot, it seems, insist that a theory be both maximally comprehensive and maximally precise, for comprehension and precision trade off against one another. We cannot, it seems, insist that every adequate explanation be physicalistic, and yet accept Woodward and Bernstein's irreducibly intentional explanation of the Watergate conspiracy. So we modify our standards of acceptability and our judgments about what accounts are actually acceptable until the two are in accord. When they mesh, both standards and judgments are acceptable.

Although our standards of acceptability are constructed on the basis of what we accept, being accepted neither entails nor is entailed by being acceptable. A system need not accommodate every relevant antecedent commitment. So not everything we accept is acceptable. And we may overlook a consideration's membership in a tenable system, or be unwilling or psychologically unable to endorse it despite its membership. So acceptability does not guarantee acceptance. Even though construction starts from what we in fact accept, the accepted and the acceptable diverge.[1]

This account is plainly pragmatic. It enables us to construe theory revision as improvement, without contending that there is a unique goal we are getting ever closer to. Just as we can recognize that a hammer is better than earlier models without measuring it against the Platonic form of a Hammer, we can recognize that a theory is better than its predecessors without taking it to be a closer approximation to the Truth. Both the hammer and the theory are tools. And there are things we want tools of each kind to do: hammers to insert nails without cracking the plaster, for example; theories to explain phenomena without resort to ad hoc expedients. In both cases, we can have good, if defeasible, reason to think the tools we have constructed have done their jobs. And both sorts of tools may be unsuitable for other, equally legitimate purposes.

A system of considered judgments in reflective equilibrium is neither absolute nor arbitrary: not absolute, for it is fallible, revisable, revocable; not arbitrary, for it is tethered to antecedent commitments. Such a system neither is nor purports to be a distortion-free reflection of a mind-independent reality. Nor is it merely an expression of our beliefs. It is rather a tool for the advancement of understanding. If it is effective, it does the job it was designed to do. That other tools do the same job equally well, or do other jobs well is no criticism of it. Neither the system nor

[1] See Catherine Z. Elgin, *Considered Judgment* (Princeton: Princeton University Press, 1997), 102–134.

the judgments it vindicates have any claim to truth or to permanent credibility. But they are, in the epistemic circumstances, reasonable and rational. They supply the considered judgments against which further revisions are to be tested.

What then are we to make of differences of opinion? Perhaps disconcertingly, the answer is: it depends. Some are unproblematic, being vindicated by alternative, equally tenable systems of thought. Thus, a cosmology that construes black holes as stars need not refute a theory that construes them as the residues of extinguished stars. Nor need a feminist interpretation of *Middlemarch* refute all other readings of the work. Other opinions genuinely clash. Then one is acceptable only if the other is not. A theory that contends that dinosaurs were warm-blooded should have the resources to discredit equally viable competitors that say that they were cold-blooded. In yet other cases, we may be hard put to tell. It is not obvious, for example, whether functionalism clashes with physicalism in the philosophy of mind. Disagreements are themselves subject to interpretation, and only under an interpretation is it determinate whether opinions actually conflict. To decide such matters, we appeal to our best theories about theories, modifying and amending them when they are inadequate to the task. Without the firm foundation that philosophical absolutes were supposed to supply or the free-floating complacency that comes from considering everything arbitrary, we have to step carefully, building and rebuilding our supports as we go. The ground under our feet is only as solid as we make it. Often it is solid enough. Not always. There are no guarantees.

Index